Made In America:

Sustainable Building Products, Materials & Methods - 2nd Edition

"We build because most human activities cannot take place outdoors. We need shelter from sun, wind, rain, and snow. We need dry, level platforms for our activities. Often we need to stack these platforms to multiply available space. On these platforms, and within our shelter, we need air that is warmer or cooler, more or less humid, than outdoors. We need less light by day, and more by night, than is offered by the natural world. We need services that provide energy, communications, and water and dispose of wastes. So, we gather materials and assemble them into the constructions we call buildings in an attempt to satisfy these needs."

Edward Allen and Joseph Iano, Authors
Fundamentals of Building Construction - Materials & Methods

Made In America:

Sustainable Building Products, Materials & Methods - 2nd Edition

Robert A. Wozniak, Jr., LEED® AP, CAPS

Associate Professor
Architectural Technology | Building Science + Sustainable Design
School of Construction and Design Technologies
Pennsylvania College of Technology
Williamsport, PA

W. R. PARKS
Hershey, PA

ISBN 13: 978-0-88493-038-9 (Paperback)
ISBN 10: 0-88493-038-6
Library of Congress Control Number: 201791184

Published by: William R. Parks
Hershey, Pennsylvania
Email: **Stanwrite@aol.com**
Web Site: **www.WRParks.com**
Twitter: **www.twitter.com/WParksPublisher**
Facebook: **www.facebook.co/wparkspublishing**

Contents:

Introduction:

Welcome! I am glad that somehow this book has found itself in your possession. As a result of not being able to find an appropriate book for one of my courses that I have taught – this resource was developed. It is a compilation, and ultimately a directory of the various American made residential and non-residential based sustainable materials that are useful to the architect, interior designer, engineer, contractor, developer, building owner, home owner, and of course, the student – who potentially is the future building / home owner and potentially a professional within the **Architectural, Engineering and Construction (AEC)** industry.

To organize the content, this book uses the **Construction Specifications Institute (CSI)** industry standard to define its chapters. This standard allows for all materials or products to be organized into various division titles for specification purposes and makes it an valuable educational opportunity. As you might know, this standard was modified from the original **16 Division** titles, to a potential of **50 Divisions** in 2004. More can be found about this at the CSI website: **http://www.csinet.org/** or by using the adjacent **QR code** .

Before we progress any further, please understand that the contact information provided, including the websites and related Quick Response (QR) codes found throughout the book are intended as a way to learn as much about the company, their product(s), Materials, Methods and other related content, as a way to compare and contrast one with another, based on the manufacturers' website content. As a building Owner, Designer/Specifier, or Contractor, such comparison is expected, so to make an educated decision to specify and install . . .*"this or that".* This video advertisement for *Kia Soul* summarizes this understanding that *"the choice is yours",* after much research to decide which product, material or method is best for you, or your clients' next project.
https://www.youtube.com/watch?v=7geyX1Er1zs 1:00

For your information, the QR Codes found in this book were created at no cost, using the website: **http://www.qrstuff.com/**

The QRStuff.com website also allows the user to download appropriate software to their mobile devices so to read the QR codes, so to take the user directly to the same website link listed near the QR code.

Ultimately, this book will focus on **Div. 02 - Div. 14,** along with **Div. 21 and22,** shown in the following list - identified in **bold letters.**

Per CSI, the Division numbers and titles are:

Division 00: Procurement & Contracting
Division 01: General Requirements
Division 02: Existing Conditions
Division 03: Concrete
Division 04: Masonry
Division 05: Metals
Division 06: Wood, Plastics and Composites
Division 07: Thermal and Moisture Protection
Division 08: Openings
Division 09: Finishes
Division 10: Specialties
Division 11: Equipment
Division 12: Furnishings
Division 13: Special Construction
Division 14: Conveying Equipment
Division 15-20 reserved for future CSI use

Division 21: Fire Suppression
Division 22: Plumbing
Division 23: Heating, Ventilation, & Air Conditioning
Division 24 reserved for future CSI use

Division 25: Integrated Automation
Division 26: Electrical
Division 27: Communications
Division 28: Electronic Safety and Security
Division 29-30 reserved for future CSI use

Division 31: Earthwork
Division 32: Exterior Improvements
Division 33: Utilities
Division 34: Transportation
Division 35: Waterway and Marine Construction
Division 36-39 reserved for future CSI use

Division 40: Process Interconnections
Division 41: Material Processing and Handling Equipment
Division 42: Process Heating, Cooling and Drying Equipment
Division 43: Process Gas and Liquid Handling, Purification and Storage Equipment
Division 44: Pollution and Waste Control Equipment

Division 45: Industry-Specific Manufacturing Equipment
Division 46: Water and Wastewater Equipment
Division 48: Electrical Power Generation

Division 49-50 reserved for future CSI use

Since the Architectural, Engineering and Construction (AEC) profession uses the CSI MasterFormat to categorize products and materials as an effective standard to efficiently specify such to a project, it would be counterproductive for this book to not use such a widely recognized system to organize *(and instruct those that are not familiar with)* the content that follows.

This resource is an intentional effort to help identify the US Manufacturers along with their sustainable products, so to make others aware that not everything comes from foreign countries, as it sometimes appears when one visits the big box stores. And, with this, hopefully such a resource could help spur the economy, and ultimately create jobs / careers. After all, if the United States cannot provide a solid employment workforce, then our nation will become a slave to others - and not be able to sustain itself.

As promoted by Diane Sawyer and David Muir of ABC News, "economist-turned-builder" **Anders Lewendal** and crew - built an ***"All-American Home"*** in Bozeman, MT using all US made products! This Made in America effort received a lot of publicity and reminded the public that not only do US manufacturers still exist, but this is what all the builders did and were capable of doing in the United States. That was of course before the surge of imports became common place in the late 1970's, and companies moved their product production to other countries - for less labor cost and potentially attractive profits for stockholders. What excited people in this 3 part News coverage was the substantiated claim of Mr. Lewendal, stating: *"if every builder used just 5% more American materials, we would create 220,000 jobs right now".* The following links and QR Codes should take you to this discussion.

YouTube: The All American Home - Part 1 link 4:20
YouTube: The All American Home - Part 2 link 4:28
YouTube: The All-American Home - Part 3 link 2:34

Part 1:

Part 2:

Part 3:

And, the following link connects you with the materials list or **'blueprint'** as mentioned and provided by Anders Lewendal.

http://abcnews.go.com/blogs/business/2011/10/how-to-build-a-made-in-america-home/

Now, as a result of watching and understanding the previously mentioned ***The All-American Home*** ABC News segments, it is up to the rest of us to follow through to stock such products in our stores, specify such for our projects, and ultimately purchase and install such American made materials and/or products, so that the end user and everyone in between can benefit from such. If the products listed in this book are not on the shelves, request such products to be stocked at the local retail and wholesale suppliers, or when possible, request samples and purchase directly from the manufacture. Also, keep in mind it is very important to not only create and use American made products, but they should be long lasting, responsibly and regionally made sustainable type quality products, that are recyclable, and are not trendy. Unfortunately, if it is **trendy**, then it is very likely that such will be torn out, discarded and replaced with the next fad. As we know, this mindset is not sustainable - even if the product that is being replaced was sustainably made!

With those products being trendy, the original ***"Story of Stuff"*** documentary, by **Annie Leonard** comes to mind. If you are not familiar, one of the topics she describes are the consumer marketing strategies known as **Perceived Obsolescence** and **Planned Obsolescence** that were developed years ago and are used today. Take a look.

http://storyofstuff.org/movies/story-of-stuff/ 21:24

How easily it is to follow the most recent design trends illustrated and marketed in magazines, on television or other means, such as tearing out the 'old' when it still has plenty of useful life left, and then replace it with a product that was made from recycled this or that, with the latest trendy color or look. Please note, that to discard items that still have useful life and replace it with the latest and greatest - falls under the topic of Perceived Obsolescence, and becomes **green washing**. It is a big eye opener to tour other countries to see their building stock that are typically much older than the age of this country.

And, what about the electronic component, device or anything that lasts only a few years at best - and is cheaper to replace than repair? This is Planned Obsolescence.

Along with the *Story of Stuff* documentary, there are various other documentaries as well. See: **http://storyofstuff.org/movies/**

Now that we have progressed to this point, it is probably best to address a few items, including: *What is* ***sustainability?*** The word gets thrown around a lot, but when it comes to our buildings; what is it? The term is relatively new, but the concept is not. Ultimately, it is the unselfish approach of being responsible for our actions and being good stewards of what has been created and provided to us. Other previous terms or phrases used in the past included: Being **Resourceful**; **Thrifty**; **Ecology; Ecological; Energy Conscious Design, Environmentalism,** which also include **Environmental Design,** and **Environmental Science.** These terms, concepts and efforts, with potential others - are the forerunners to **Sustainability** before and after the 1970's first oil crisis.

Then in 1987, **The United Nations' Brundtland Commission** most recognized definition became:

> *"Sustainable development is development that meets the needs of the present without compromising the ability of future generations to meet their own needs."*

As a good example to the above quote, the **Seventh Generation Company**, a company that makes various plant based household products, gets its Mission and Values from the 16th century Iroquois Confederacy where they created the Great Law that instructs that:

> *"in our every deliberation, we must consider the impact of our decisions on the next seven generations."*

Take a moment to look over and consider their products. See: **http://www.seventhgeneration.com/**

Then ultimately, for a product to be considered sustainable, a **third party analysis** should be done, and is done to critique the raw materials, the manufacturing process, the total embodied energy

used including the manufacturing and transportation costs. Other considerations include product longevity and its ability to decompose naturally, be reused, or be recycled after its original use. **Inhabitat** offers an article known as: ***Green Building Labels 101: An Inhabitat Guide to Third Party Environmental Certifications.*** This lists the better-known 3rd Party Certifying organizations and discusses what are 1st and 2nd Parties as well - ultimately to promote product transparency. See:
http://inhabitat.com/green-building-labels-101-an-inhabitat-guide-to-third%C2%AD-party-environmental-certifications/

The Redwoods Group Foundation offers the below ***Green Plus*** website that also discusses 3rd Party Certification further with additional considerations. See:
http://gogreenplus.org/nuts-and-bolts-guide/planet-nuts-and-bolts-guide/sustainable-purchasing/product-sustainability-certifications/

What is ultimately needed and is finally coming to pass is when there is a demand for American made, long lasting, responsibly made, sustainable products, it becomes a win-win opportunity. The American companies thrive on the quality reputations - creating and maintaining a sustainable workforce (from manufacturing to installation), and then the end user ultimately benefits as well with such products and materials!

But, again, before we go any further, ***What Makes A Product Green?*** The following list of 'high-performance attributes' is elaborated upon in an article by the same name at **BuildingGreen.com**, asking - *Was it?; Does it?* or *Will it? . . .*

- be made with Salvaged, Recycled, or Agricultural Waste
- Conserve Natural Resources
- Avoid Toxic or Other Emissions
- Save Energy or Water
- Contribute to a Safe, Healthy Built Environment

https://www.buildinggreen.com/feature/building-materials-what-makes-product-green

Yet, the simplest definition is:

"The greenest material is one that you don't use."

Michelle Kaufmann, Architect & Author; *The Not So Big House*

Along with all that has been mentioned thus far, and all that will be presented in the rest of the book, **Building Science** is important. This term and its professional practice encourages us to design holistically so to assemble all the materials, products, the building's orientation to the site, and all the other design considerations - so to consider the entire building, its site, its components, systems, the end users, and their needs, while collaborating with all stakeholders possible, thus recognizing and understanding that . . .

> *"**Building Science** is the cross disciplinary collection of knowledge and experience required to understand and predict many aspects of the behaviour (performance) of buildings and their systems, specifically including: Durability, comfort, energy, environmental separation, Indoor Air Quality, Acoustics, lighting, economics and constructability".*
>
> Prof. John Straub, RDH / Building Science Labs

Building Science Labs link

Building Science Corp link

Disclaimer:

The intent of this book is to provide a resource of sustainable materials that are manufactured within the USA, whose contact information can be found. The goal was to list these USA manufacturing companies, not their distributors.

This resource guide is simply a directory that strives to organize and communicate as many residential and non-residential based US Made sustainable products & materials as possible . . . by the men and women who creatively have designed & manufactured them, with the goal to share these with the designer, contractor, building owner and others, for the next ***Alterations, Additions*** or ***New**** construction project. (* = Actual terms for typical types of projects addressed and defined in the Building Code.)

None of the products, materials, manufacturers, organizations, agencies, or individuals mentioned within are endorsed by the author or the publisher of this book. This book is merely a resource allowing the user to locate, compare and seek further information from the contact information provided within, so to make an informed decision, so to determine which American Made, green product is best for the desired application. By accessing the information, it is important that any Terms & Conditions of the manufacturer are adhered to. The related websites are meant to be referenced to obtain the latest information. Over time, not only will the website be updated, but the actual link may have been changed or no longer accessible. Of course there is no control over this. ☹

Along with the materials and products, there is some additional content, cross referencing, editorial comments and some Building Code references that are meant to help the reader integrate applicable related minimum requirements.

With the following, if you know of a product, material, or organization, whose specific information provided should be added or updated, or one that is requested to be removed in a future re-printing – please contact the author at: *RobWozJr@gmail.com*

End of Introduction

Space for notes:

DIVISION 02 - Existing Conditions

This CSI **Division 02** contains specific resources that relate to the existing conditions that are possible at the building site. This could include any possible concerns related to any potential existing buildings on site, along with any possible concerns that could be below and/or above grade, including but not limited to property line disputes, building setbacks, erosion control, various assessment issues, and site, water or facility/building remediation needs. As stated previously, **Division 00** Procurement & Contracting, and **Division 01** General Requirements, are not elaborated upon, as they typically are not material or product based divisions. **Division 02** however has some products that follow, but Division 02 tends to be more professional service oriented. In those cases, through various references, it is hoped that quality professionals can be found as close as possible to the project site - so to minimize costs.

See International Building Code (IBC) *Section* ***1803*** - ***Geotechnical Investigations*** for related information.

Use this website for *PremiumACCESS or* free *PublicACCESS* viewing of ***I-Codes*** *and* ***ICC Standards***: **https://codes.iccsafe.org/**

02 00 00 - Existing Conditions

This section typically applies to services, such as Maintenance of the Site and/or Building. It applies to the various existing items found on site that should be addressed prior to any new work. See the following **Division 02** Sections for specific Existing Conditions Section headings.

02 05 19 - Geosynthetics for Existing Conditions

With the **Municipal Separate Storm Sewer System (MS4)** emphasis per the *EPA MS4* ***Stormwater Program Overview***

related link, **http://www3.epa.gov/npdes/pubs/fact2-0.pdf** - this attempts to retain the agricultural pollutants and construction site sediment that is transferred via drainage systems during ordinary and more severe rain storms that provide run-off that travels along county roads, municipal streets, curbs and gutters to ditches and man-made catch basins, storm drains and **bio-swales**.

Geosynthetics are important when properly used and placed, as they can provide **soil stabilization** and prevent unnecessary erosion and related repair costs. Geosynthetics include environmentally conscious replacement for stone, sand, and asphalt such as: Geocells, landscape fabric, frost blanket, grass reinforcement mesh, land protection, engineered turf, beach access matting, root barrier, and porous pavement applications.

It needs to be said however, if man & woman intervened along the Colorado River by applying this same emphasis years ago to control soil erosion, would we have the Grand Canyon today?! If not, this would mean that we would only have six *Wonders of the Natural World*. Something to think about.

https://en.wikipedia.org/wiki/Seven_Natural_Wonders_(CNN)

Erosion has been part of our planet for ages. Consider the sandy beaches and sand boxes that we all enjoy. But, when necessary, there are always sandbags, and other various sustainable products to try to tame the results of weather, including:

- **TYPAR Geosynthetics**
  Fiberweb, Inc.
  70 Old Hickory Blvd.
  Old Hickory, TN 37138
  Phone: 800-541-5519

  www.typargeosynthetics.com

The **United States Green Building Council (USGBC)** addresses erosion as a **Sustainable Sites (SS)** prerequisite requirement for when **Leadership in Energy and Environmental Design (LEED)** Certification is pursued. **http://www.usgbc.org/education/sessions/leed-construction-phase-lesson-1-ss-prerequisite-construction-activity-pollutio-0**
The above follows the **Environmental Protection Agency (EPA) National Pollutant Discharge**

Elimination System (NPDES) program is a result of the Federal **Clean Water Act**.

02 20 00 - Assessment

This section is typically not materials, but services, such as: Site or Building Surveys, Studies and Assessments. For such professional services, it is suggested to obtain referrals for the companies or firms that provide the most reliable services in your area. When working with older, potentially historic buildings, including those on the **National Register of Historic Places**, review each of the National Park Service **Preservation Briefs**, and specifically ***Brief 35***, listed below . . .

- **Understanding Old Buildings: The Process of Architectural Investigation**
  National Park Service
  US Department of the Interior
  Technical Preservation Services
  Preservation Brief 35

  https://www.nps.gov/tps/how-to-preserve/briefs.htm

Of course, Architectural Investigation or other assessments, may require the services of a Licensed / Registered Architect or Professional Engineer (PE), that has experience in Civil, Structural, Mechanical, Electrical or Plumbing, specific to the project State/Jurisdiction.

Continuing on the topic of Assessments, one that has become very popular today, is a building **Energy Audit.** Such a building study is typically very successful in identifying where the residence or non-residential building could save on energy costs for the owner or tenant, while optimizing the living or working environment. Energy Audits are profitable for both new and existing construction. One reputable organization that trains and certifies individuals to perform Energy Audits is:

- **Building Performance Institute (BPI)**
  107 Hermes Road, Suite 210
  Malta, New York 12020
  Phone: 877-274-1274 : 518-899-2727
  FAX: 866-777-1274 : 518-899-1622

  http://www.bpi.org/about-us

A buildings' construction could be built initially or improved by appropriate renovations to meet a variety of current **energy standards**. See a listing and descriptions of the various energy based standards at:

- **Whole Building Design Guide (WBDG)**
  National Institute of Building Sciences (NIBS)
  1090 Vermont Avenue NW, Suite 700
  Washington, DC 20005
  Phone: 202-289-7800
  Fax 202-289-1092

  http://www.wbdg.org/resources/gbs.php

Along with . . .

- **International Energy Conservation Code (IECC)**
- **ICC Performance Code (ICCPC)**
- **International Green Construction Code (IgCC)**
  Note: As referenced earlier, Free ICC Code access is found at **https://codes.iccsafe.org/**

Another assessment for both the existing building and its site typically will include evaluating how *accessible* is it? **Universal Design** has taken on various names in the past, including but not limited to **Inclusive, Barrier-Free** and **Handicapped Accessible Design.** Gaining access to buildings took center stage in 1990 with the passing of the **American with Disabilities Act (ADA).** This is commonly understood as an extension of the *Civil Rights Act* of 1964 – prohibiting discrimination, and allowing access to buildings of public accommodation operated by the US Government or private entities. In other words, this is not code, but law. The ADA standards are issued by the **US Department of Justice (DOJ)** and the **US Department of Transportation (DOT)**. Depending on the facility and application depends on if the DOJ or DOT standards apply.

- **ADA Standards for Accessible Design**
  United States Department of Justice
  Civil Rights Division
  950 Pennsylvania Avenue, NW
  Civil Rights Division
  Disability Rights Section - NYA
  Washington, D.C. 20530
  Phone (202) 307-0663
  Fax: (202) 307-11097

  http://www.ada.gov/ada_intro.htm

The US Department of Justice (DOJ) creates standards for:

- Assembly Areas

- Detention and Correctional Facilities
- Housing at Places of Education
- Medical Care Facilities
- Places of Lodging
- Residential Dwelling Units *(Gov't subsidized / HUD)*
- Social Service Center Establishments

▪ **US Department of Transportation**

Federal Transit Administration
East Building
1200 New Jersey Ave., SE
Washington, DC 20590
Phone: 866-377-8642
TTY: 800-977-8339

http://www.fta.dot.gov/civilrights/12325_3884.html

The US Department of Transportation (DOT) creates standards for public transportation, including:

- Location of Accessible Routes
- Detectable Warnings on Curb Ramps
- Bus Boarding and Alighting (descending) Areas
- Rail Station Platforms

The **United States Access Board (ADAAG)** has more clarification and specifics of both DOJ & DOT above.

▪ **United States Access Board**

US Access Board Accessibility Guidelines (ADAAG)
1331 F Street NW, Suite 1000
Washington, DC 2004-1111
Phone: 202-272-0080 or 800-872-2253
TTY: 202-272-0082 or 800-993-2822
Fax: 202-272-0081

https://www.access-board.gov/

Along with the above, the **International Code Council (ICC) / ANSI** provide the electronic or print publication:

▪ **ICC/ANSI A117.1 - Accessible and Useable Buildings and Facilities**

ICC Store
USA: 1-800-786-4452
International: 708-799-2300, ext. 33801
order@iccsafe.org

https://codes.iccsafe.org/ - *Use ICC Standards Link*

It would be important to access and be familiarized with the ICC/ANSI A117.1 Standards. It should also be noted that the **ICC International Building Code** (IBC) has Chapter 34 – dedicated to ***Existing Buildings***, along with an entire **International Existing Building Code** (IEBC). These address how and what is done to accommodate various accessibility concerns with an existing building, with the emphasis on getting someone from "point A" to "point B" independently, with dignity.

Along with this, keep in mind that there is the **ADA 20% Disproportionality Rule** as it relates to Existing Buildings, where only 20% of the total cost is required to address accessibility beyond the *'Primary Function Area'.* See more information, including *'Prioritization'*, at: **https://www.access-board.gov/guidelines-and-standards/buildings-and-sites/about-the-ada-standards/guide-to-the-ada-standards/chapter-2-alterations-and-additions**

Free Public Code electronic access:

*As indicated previously . . . the IBC, IRC, along with the **ICC/ANSI A117.1** - and all the other ICC and State codes can be accessed FREE via:* **https://codes.iccsafe.org/** This site also provides an interactive United States map that communicates which codes have been adopted for each particular State.

Remember, if we don't understand the minimum standards of the building code (or zoning requirements) for a particular location, we are simply wasting time, and trying to fool ourselves. Such requirements create a bench mark for what should or should not be. As the designer or contractor, introduce yourself to the local **Authority Having Jurisdiction (AHJ)** (or otherwise known as the Building or Zoning Inspector), so to best understand his or her expectations and code interpretations.

Typically, private single-family (owner-occupied, non-governmental tax subsidized) residences are not required to accommodate the **Americans with Disabilities Act**. With the private single-family (owner occupied, non-governmental tax subsidized) residence, homeowners typically when faced with a family member who has limitations as a result from Birth, Accident, Disease or Age, are forced to either "make do" because of financial or design limitations; possibly move temporarily when renovations are being made; or make a permanent move to a home that offers better accommodations. Even if built using the latest sustainable products and materials while meeting or exceeding a selected energy standard, if the occupant or owner no longer can gain access or move about independently in his or her own home or place of worship - how is this sustainable? What a crime! When designing any building, including residences and places of worship, whole-building design should include site integration, Indoor Air Quality, sustainable materials and product integration, and related building assemblies are very important, but accommodations should also be integrated to make a building accessible for all - which then ultimately will sustain the occupants for the long-term.

Making alterations to an existing building to accommodate accessibility may be possible if space and/or structural loading permits
, but it typically comes with a price tag. Such alterations typically include:

- Accessible Entry (Zero Step / Step Free entrance)
- Accessible Horizontal Circulation (within the building)
- Accessible Vertical Circulation (Small elevator, LULA or stair lift)
- Accessible Doors / Doorways (Since only one exterior entrance door is required by Code to be 36" wide, all others may need to be modified. Accessibility requires 32" min. clear openings, measured from the face of the open door to the door jamb. (Ultimately, to accommodate a wheelchair). A 36" wide door is the only one that provides such clearance, unless offset hinges are integrated.)
- Accessible Electrical Controls (36" - 42" AFF for switches & HVAC controls, and 18" - 24" AFF for receptacles / outlets.)
- Reinforced Walls (for future grab bar / railing installation in the bathrooms, hallways, steps and other possible locations.)
- Accessible Kitchens and Bathrooms. (Providing various height counters and other considerations to accommodate various heights and abilities of the occupants and visitors.)
- Accessible Bedroom (First floor bedroom is typically standard feature in a Ranch or Cape Cod style residence.)

While anticipating the above alterations, it is important to keep in mind that there probably will be a need to replace and/or patch to match all floor, wall and ceiling finishes that were affected when the door was widened, or any other work done during renovations! These renovation costs could have been eliminated or minimized if the original design considered and anticipated the **"Sustain-Ability"** design approach. These design considerations address long term use and accommodations for all people, *(working, residing and/or visiting)* in the present, or the future. For some reason it is not being emphasized elsewhere, but **Universal Design** is the ultimate in **Sustain*Able* Design!**

When designing new, or making alterations to an existing building so to make it more energy efficient

. . . if a building meets the minimum or most stringent sustainable criteria, (be it per the International Energy Code, Energy Star, National Green Building Standard (NGBS) LEED, Passive House (PHIUS) or another standard), yet, does not address Universal Design (and the anticipated future needs of all occupants) . . . what good is it!?! It does not accommodate/sustain all people.

Along with the above topic of Assessment and Building Codes also brings to mind what will be the Building Use / Occupancy? Here, we get into the topic of **Occupancy Groups**. And, whether it is an existing building or new, the space or entire building that is being addressed will need to be assigned to one of the ten (10) Occupancy (or Use) Groups, (if it is not a One or Two Family Dwelling or Townhouse). Or, maybe it will be divided into two or more - known as Mixed Use Occupancy. Either way, before you can analyze the code for your specific use and requirements, this needs to be determined. Use the **Free Public Code electronic access** (discussed previously) to find the International Building Code (IBC), **Chapter 3** - *Use & Occupancy Classification.*

Some additional related resources are:

- *See* ***Division 14 -*** *Conveying Equipment, for related Universal Design products and information.*

- ***International Building Code*** *(IBC) - Chapter 11, Accessibility.*
(It should be noted that churches built using the 2015 IBC, Section ***1102.8, Areas in Places of Religious Worship****, are now required to have accessible areas, except where noted in Section 1102.8.*
https://codes.iccsafe.org/public/document/code/542/9678513

- ***National Association of Homebuilders (NAHB) - Certified Aging In Place Specialist (CAPS)*** provides training and certification geared towards the important and ever growing residential Aging-In-Place / Universal Design demand, as the largest current Baby Boomer (born between 1946 - 1964) generation retires..

02 30 00 - Subsurface Investigation

For subsurface investigation, contact an experienced local Registered Architect or Engineer.

02 40 00 - Demolition and Structure Moving

Contact the below for a reputable contractor:

- **National Demolition Association**
  16 N. Franklin Street, Ste. 203
  Doylestown, PA
  Phone: 800-541-2412
  Fax: 215-348-8422

  http://www.demolitionassociation.com/

Demolition should be the last option for any structure. See related topics:

- *Preservation, rehabilitation, restoration or reconstruction under CSI Section* ***02 80 00 - Facility Remediation***

If the building cannot be saved from demolition, or it cannot be moved then the building materials should be salvaged. See the Building Materials Reuse Assoc. under Section 02 80 00 - Facility Remediation

- **Int'l Association of Structural Movers** (IASM)
  PO Box 2104
  Neenah WI 54956-2104
  Phone: 803-951-9304
  Fax: 920-486-1519

  http://www.iasm.org/

If the building or structure cannot be kept on its original site, then it should be moved in lieu of demolition.

02 50 00 - Site Remediation

Contact an experienced local Registered Architect, Engineer or an appropriate professional for any of the issues or concerns related to an existing site that may include these common issues or concerns, listed alphabetically:

- **Accessibility**
  Similar to what was discussed previously, but now with the emphasis on the existing building site. With the site, is there accessible parking, accessible pathways and other various related **handicapped accommodations** on site that should be addressed per ADA and Universal Design requirements?

- **Brownfield**
  Is there any trace of **harmful substances found on site** (on the ground surface or in the water), including, but not limited to chemical based cleaners, gasoline, oil, lead, CCA Pressure Treated Lumber, asbestos, fertilizers, weed killers, naturally occurring Radon, or a large scale industrial waste site?

  If there are any **harmful substances buried** below, 'downstream', or leaching from neighboring property, such as from a small automotive repair shop, to a large chemical plant, to nuclear dump sites? Is there an unusual amount of disease, birth defects or illness in the area? Or, what about that nice open space with walking trails, or the public park that unknowingly is tainted from being a the past dump site? If so, such an area could be an undocumented 'Brownfield'. See the waste lands that are documented . . .

  http://projects.wsj.com/waste-lands/

  See related *CSI Section* ***02 60 00 - Contaminated Site Material Removal***

- **Buried Utilities**
  Are there any **buried utilities** on site? Any professional excavator, building owner or homeowner <u>is required</u> to always ***"Call (811) before you dig!"*** This same 3-digit number applies to all digging, in all States! **http://call811.com/**

- **Diseased Trees or Vegetation**
  Are there **diseased trees or vegetation** that need to be removed to help stop the disease from spreading? One source for this is Arbor Day:

https://www.arborday.org/trees/tips/identifying-pests-and-diseases.cfm

- **Erosion & Sediment Control**
  Is there a need for **erosion & sediment control** as communicated by the *United States Environmental Protection Agency (EPA)?*

- **Flood Plain**
  Is the site within a **flood plain** as indicated on the *United States Federal Emergency Management Agency (FEMA) Flood Maps*?

- **High Water Table or Wetlands**
  Is there a **high water table** or **wetlands** on the property, as defined by the *United States Environmental Protection Agency (EPA)?*

- **Sinkholes**
  Is the lot/acreage in a region prone to **sinkholes** similar to Central Florida, and/or studied by the *United States Geological Society (USGS) – Water Science School?*

- **Soil Types**
  Are the **soil type(s)** appropriate for structural bearing, drainage, septic system percolation, growing vegetation and trees? If for structural reasons, requested soil boring test locations are commonly indicated on the site survey and then the tests are typically performed during Schematic Design phase of a project.

- **Other concerns include:**
  Toxic substances are found in various pesticides, cleaners, detergents, paints, and even some imported goods.

  With sustaining human life, we are faced with **Genetically Modified Organisms (GMOs)** as found in various commercialized grown **Genetically Engineered (GE)** plants and animal feed.

  GMOs are not typically associated with Site Remediation, but the absence of them can be an important part of soil, animal and human sustainability. What will be the results of this scientific experiment of GMO based goods and food products that are being manufactured, ingested or even burned as fuel?!

With GMOs, what do you know? Below are a few links:

- The *Non-GMO Project*: (Including GMO Facts) **http://www.nongmoproject.org/learn-more/**

- The *GMO-OMG* film, by Jeremy Seifert. **http://www.gmofilm.com/**

- The common GMO based plants grown in the USA, were adopted as recently as 1996. **http://time.com/3840073/gmo-food-charts/**

- GMO Free Regions: Several countries around the globe understand that GMOs as harmful, and as a result ban them. Unfortunately, the United

States is not one of them.

http://www.gmo-free-regions.org/gmo-free-regions.html

- The list of countries that require food labels to identify if the item contains GMOs. Unfortunately again, the US does not require it, thus prohibiting transparency.

http://www.justlabelit.org/right-to-know-center/labeling-around-the-world/

- This link from the US Food and Drug Administration (FDA) communicates about food from Genetically Engineered (GE) Plants.

https://www.fda.gov/food/ingredientspackaginglabeling/geplants/default.htm

- Five major US food companies announce voluntary GMO labeling. **http://www.gmoinside.org/five-major-food-companies-announce-gmo-labeling-what-should-we-expect-at-the-grocery-store/**

- "Breakfast is *not* a science experiment!" Here is an opportunity to send an online form letter to

General Mills, to request that GMOs be removed from Honey Nut Cheerios. (Original Cheerios used to be made with GMOs, but with a similar campaign by GMO Insider.org in 2012, they were removed.)

http://action.greenamerica.org/p/dia/action/public/?action_KEY=9239

Additional information about the campaign to remove GMOs from Honey Nut Cheerios.

http://www.gmoinside.org/portfolio_page/cheerios/

The above are just a few of the many related links about GMOs and GE plants and food products. Search others to form your own opinion, including *CSI Section* ***07 21 29 - Spray Foam Insulation.***

- **Bee Health** -
  Although not commonly associated with Site Remediation, it does effect and relate to the pollination of food producing plants for our nation and world and is important concern for everyone. Bee health has become an issue since 2006 and is commonly referred to **Colony Collapse Disorder (CCD).** A cell phone study was done that offers possible connection to the problem and the electromagnetic confusion it offers bees migrating back to their hives.
  See related article & video at:

  http://www.cnn.com/2010/WORLD/europe/06/30/bee.decline.mobile.phones/

  Bee health is not typically associated with Site Remediation, however the absence of bees would be detrimental to pollination and related food growth.

02 60 00 - Contaminated Site Material Removal

If a project site contains contaminated soils, it could be considered a **brown field.** One method that minimizes expenses is the proven **cement based soil solidification / stabilization** technique that prevent relocating soils (and related costs and the underlying problem) to a new location. A related video, can be found at :

http://www.youtube.com/watch?v=aOvo8axDTaw 9:35

The above Soil Stabilization video is by:

- **Ground Up Road Construction**
  Sumner, WA
  Phone: 253.891.1413
  Fax: 253.891.1425

  http://grndup.com/

The EPA provides a list of technologies used to clean up contaminated sites. See:
http://www.epa.gov/tio/download/remed/ss_sfund.pdf

- **Portland Cement Association** (PCA)
  500 New Jersey Ave., NW 7th Floor
  Washington DC 20001-2066
  Phone: 202-408-9494
  Fax: 202-408-0871

  http://www.cement.org/

02 70 00 - Water Remediation

Water Remediation commonly refers to the processes that are involved to restore ground water quality. Instead of simply sending the storm water directly down a storm drain to a nearby body of water, a few valuable techniques that provide this water a chance to be cleansed while also recharging the aquifer, are:

- Retention Basins
- Dry wells
- Bioswales

Contact the following two resources to obtain referrals for the companies or firms that provide the most reliable services in your area.

- **National Ground Water Assoc.** (NGWA)
  601 Dempsey Rd.
  Westerville, OH 43081
  Phone: 800-551-7379
  (614 898.7791 outside the United States)
  Fax: 614 898.7786
  Email: ngwa@ngwa.org

  http://www.ngwa.org/Pages/default.aspx

It is noted that NGWA's website communicates *"After the Drought"*. As we know, droughts historically happen in cycles. There were several references to droughts & famines recorded in the Old Testament. One of the most prominent was during the Egyptian Era, where a seven year drought was documented

after seven years of abundance. See: *Genesis Chapter 41, at link below.* **https://www.biblegateway.com/passage/?search=Genesis+41&version=NIV**

- **Ground Water and Ecosystem Restoration Information Center** (GWERIC)
  P. O. Box 1198
  Ada, Oklahoma 74821-1198
  Phone: 580-436-8502
  Fax: 5 80-436-8503

  https://www.fedcenter.gov/Bookmarks/index.cfm?id=9981&pge_prg_id=23417&pge_id=1869

With the 2010 Gulf of Mexico, BP Macondo Oil Well disaster, it is said that natural oil eating microbes where potentially more effective in cleaning up the oil, compared to the 1.8 million gallons of dispersants. It is also learned that the ecosystem was restored much quicker than anticipated, again thanks to the microbes, and the chemicals used are "capable of injuring people and wildlife". Man's intervention to rectify, can make things worse. See the following link and search for other related articles:

https://www.washingtonpost.com/news/morning-mix/wp/2015/04/07/study-suggests-chemical-used-in-bp-oil-spill-cleanup-capable-of-injuring-people-and-wildlife/?utm_term=.f0d5b66f8f36

*See **Division 22 – Plumbing**, for additional information regarding water.*

02 80 00 - Facility Remediation

As it relates to existing buildings or structures, the design profession has shown and communicated that the ***'greenest'* building is the one that already exists.** Such a statement considers all the raw materials, **embodied energy**, and craftsmanship that were used to fabricate and install each and every material and product that went into constructing the existing building or structure. Such buildings or structures help to define a <u>sense of place and time</u> that are unique to each community. **Restoration or adaptive re-use** is encouraged in lieu of demolition and disposal. However, such buildings or structures may have **existing conditions** that require remediation. It is understood that sometimes as a result of a **Feasibility Study** that the “economies of scale” may favor demolition. This is especially true if the structure is burned out from a fire or severely damaged from earthquakes or the like. Depending on the degree of work involved, the **National Park Service, *Technical Preservation Services*** provides various information including *The Standards, providing Four Approaches* to *the Treatment of Historic Properties*, be it **preservation, rehabilitation, restoration** or **reconstruction**.

Find more information at:

- **National Park Service** (NPS)
  Secretary of the Interior's Standards
  Technical Preservation Services
  1849 C. Street, NW
  Washington, DC 20240
  Phone: 202.513.7270
  Email: NPS_TPS@nps.gov
  http://www.nps.gov/tps/standards.htm

For direct access to the valuable **NPS Preservation Briefs**, for "how-to" procedures on a variety of topics, see: **https://www.nps.gov/TPS/HOW-TO-PRESERVE/BRIEFS.HTM**

Some additional related resources are:

- **National Trust for Historic Preservation**
  The Watergate Office Building
  2600 Virginia Ave. NW, Ste. 1100
  Washington, DC 20037
  Phone: 202.588.6000
  Toll Free: 800.944.6847
  http://www.preservationnation.org/

- **PreserveNET**
  Cornell University
  Ithaca, NY
  http://www.preservenet.cornell.edu/index.php

- **National Council for Preservation Education** (NCPE)
  http://www.ncpe.us/

On a side note, with the above resources, many times there are **Internship, Scholarships,** and **Career employment opportunities** available. And, it is important to remember, that any of the companies and organizations listed throughout this book would also be good for summer or full time employment. Take a look. You just never know!

It should be noted that if a building, for whatever reason cannot be restored, then the various valuable components should be parted out and salvaged, rather than demolished and taken to the municipal landfill. The **Building Materials Reuse Association (BMRA)** offers various resources including a directory of companies that stock various materials that were saved from a wrecking ball and now are available for reuse. See: **https://bmra.org/bmra/**

Depending on the **scope** of work involved, (in other words, all the work desired or required), including, *location, and other factors),* to build new at times can be less expensive, as there usually are unforeseen items to address when working with an existing structure – thus with and existing building, in many instances, the owner is encouraged to have a **contingency fund** for unexpected issues and expenses that may appear. However, in many instances, an existing structure is connected to an existing infrastructure, and offers character, history, quality of materials, and craftsmanship that are hard to replace. And, depending on the **scope** of the project, location and the other factors – in many cases it proves itself profitable to work with the existing structure.

Again, depending on the scope, one option to keep in mind is the possible financial incentives offered by each State and regions. Learn about each of the current incentives using *the* ***D****atabase of* ***S****tate* ***I****ncentives for* ***R****enewables &* ***E****fficiency* ***(DSIRE)****. See:* **http://www.dsireusa.org/**

And, before any work is started, many times a non-biased **Feasibility Study** is completed by an Architectural firm, to determine anticipated costs and potential alternatives, for the work involved.

Continuing the **02 80 00 - Facility Remediation** topic, the various items that follow could add substantial costs to a project or end up being financially impossible to correct. They represent various topics, issues or concerns that could be found on the exterior and/or within the interior of the structure. These common issues or concerns are listed <u>alphabetically</u>:

- **Accessibility**

As mentioned previously, since this Section focuses on Facilities, we need to ask, once in the building, does the existing building or structure meet the various related **handicapped accommodations** that need to be addressed by law, per the Americans with Disabilities Act (ADA)? *See CSI Section **02 20 00** - **Assessment** for additional information.*

- **Asbestos**

Per the United States Environmental Protection Agency (EPA), Asbestos is commonly found in Vermiculite insulation, some vinyl (asbestos) flooring tiles (VAT); the backing on sheet vinyl flooring; some roofing and exterior siding; textured paint and patching compounds applied to walls and ceilings; some heat resistant surfaces; some hot water and steam pipe insulation; some oil and coal furnace door gaskets; some heat resistant fabrics; and some automobile clutches and breaks, to name a few. See"
http://www2.epa.gov/asbestos

The remediation process for asbestos commonly includes *proper* removal and disposal, or appropriate and safe **encapsulation**.

A few encapsulation products include:

- **Asbestos Safe**
  Global Encasement, Inc.
  132 – 32nd St.
  Union City, NJ
  Phone: 800-266-03882
  http://www.encasement.com

- **Encapsulguard**
  Hy-Tech –
  P.O. Box 216,
  Melbourne, FL 32902
  Fax: (321) 984-1022
  http://hytechsales.com/encapsulation.html

- **Carbon Monoxide (CO)**

A dangerous odorless gas – that causes harmful health effects, including death! The **Clean Air Act** of 1970 was amended in 1990. It requires the EPA to set National Ambient Air Quality

Standards. See: **https://www.epa.gov/indoor-air-quality-iaq/carbon-monoxides-impact-indoor-air-quality**

Back drafts from combustion fueled appliances and/or idling vehicles in poorly ventilated spaces are common causes of CO poisoning, that can lead to death.

At the above EPA website, there is a valuable link, labeled as: *"About Carbon Monoxide Detectors".*

- **Carpenter Ants**

Carpenter ants typically burrow into wood that is prone to moisture to establish their nests. Learn more at Pestworld.org, the National Pest Management Association website:
http://www.pestworld.org/multimedia-center/pest-tv/educational/carpenter-ants/ (includes video)

- **CCA Pressure Treated Lumber**

Typically made from **Chromated Copper Arsenate (CCA),** this treated wood was used commonly for decks, playground equipment, raised vegetable garden beds, fences and outdoor furniture, where unfortunately, people (and animals) came in contact to it, and had various related health claims. It was marketed under multiple private company names that injected CCA into wood typically with a 30yr. guarantee. Unfortunately, various illnesses were the result of exposure to it. Per the EPA, effective December 31, 2003, only with certain exceptions, CCA cannot be made or used for residential uses. See this link for the EPA "Overview of Wood Preservative Chemicals".
https://www.epa.gov/ingredients-used-pesticide-products/overview-wood-preservative-chemicals#reregistration

The remediation process for CCA lumber commonly includes proper **removal and disposal**, *or* appropriate and safe **encapsulation**.
Never burn it, and wash thoroughly after handling it!

One product that can be used to encapsulate is:
Encapsulguard
Hy-Tech –
P.O. Box 216,
Melbourne, FL 32902
Fax: (321) 984-1022
http://hytechsales.com/encapsulation.html

As of a result to the ban on CCA for residential uses, **Alkaline Copper Quaternary (ACQ)** Pressure Treated Lumber is one alternative that is commonly sold.

Other alternatives to pressure treated lumber are:

- Cedar
- Composite Lumber
- Redwood
- Thermally Treated Lumber
- *and*, some exotic lumber, such as: Ipe, Cumaru, Tigerwood, and others. However, these are not grown in the USA.

*See **Division 06** – Wood, Plastics and Composites, for more information*

- **Deplorable Conditions**

The Building Code's main purpose is to protect public ***Health, Safety*** *and* ***Welfare*** - so to prevent the problems and illnesses that occurred in the 19th c. with overcrowding tenement housing. And yet, for some reason, in the 21st c., there are still owner occupied buildings and/or apartments that make the nightly news that are not fit for habitation - because of health and/or safety concerns. Or worse, they catch fire and cause fatalities. It is everyone's responsibility to speak up to remedy a situation.

- **Lead Paint**

Per the EPA, any structure that was built prior to 1978 may contain paint that was manufactured with lead. And, as of as of April 22, 2010 the EPA Lead Paint Law took effect. **http://www2.epa.gov/lead**

The EPA's **Lead Renovation, Repair and Painting Rule** (RRP Rule) requires that a professional painter who will perform renovation, repair, and painting projects that disturb lead-based paint in homes, child care facilities and pre-schools *(built before 1978)* shall be certified by EPA.

http://www2.epa.gov/lead/renovation-repair-and-painting-program

The remediation process for lead paint commonly includes proper **removal and disposal**, or appropriate and safe **encapsulation**.

For encapsulation applications, a few products include:

- **Encapsulguard**
  Hy-Tech –
  P.O. Box 216,
  Melbourne, FL 32902
  Fax: (321) 984-1022

  http://hytechsales.com/encapsulation.html

- **Lead Lock**
  Global Encasement, Inc.
  132 – 32nd St.
  Union City, NJ
  Phone: 800-266-03882

  www.encasement.com

Lead & other concerns: *Some additional information*

- **Appropriate Methods of Reducing Lead-Paint Hazards In Historic Housing**
  0National Park Service

US Department of the Interior
Technical Preservation Services
Preservation Brief 37

https://www.nps.gov/tps/how-to-preserve/briefs.htm

- **Lead from Plumbing Fixtures**
  Plumbing fixtures manufactured prior to January 4, 2014 have higher concentrations of lead. As a result of the ***Reduction of Lead in Drinking Water Act,*** pipes, pipe fittings, fittings or fixtures will have a significant reduction 36 months after being signed into law. The deadline would have been January 4, 2014. See: ***https://www.gpo.gov/fdsys/pkg/PLAW-111publ380/pdf/PLAW-111publ380.pdf***

So, any plumbing manufactured or installed prior to this would probably have more lead concentrations.

See: *Get the Lead Out Plumbing Consortium*: **http://www.pmmag.com/keywords/3910-get-the-lead-out-plumbing-consortium**

It is recommended by the EPA to: *"Flush water outlets used for drinking or food preparation".*

http://www2.epa.gov/lead/learn-about-lead#exposed

So, if a faucet is dripping, this technically is a good thing.

See ***Division 22 – Plumbing****, for related information*

- **Mercury from CFL's**

  Because **Compact Fluorescent Light (CFL) bulbs** contain Mercury, they have been considered a toxic time bomb! We are replacing thermostats that contain Mercury, so why is Mercury used in light bulb manufacturing? See the EPA's website on how to "clean up a broken CFL", if it breaks.

  http://www2.epa.gov/cfl

  Yes, CFLs use less electricity as artificial light than an incandescent bulb, but at what expense to all living things? Why were CFLs permitted to be manufactured and sold using Mercury? A Government mandate to phase out the old and usher in the new resulted in closing US manufactured incandescent factories, allowing the majority of CFLs to be manufactured in China. This Government involvement was unprecedented.

  With the fairly rapid advancement of **LEDs** in recent years, we no longer have to fear Mercury, but unfortunately they do *"contain lead, arsenic and a dozen other potentially dangerous substances".*

  https://www.scientificamerican.com/article/led-lightbulb-concerns/

  The good 'ole *"simple* ***incandescent light bulb*** *is still the cleanest, most non-toxic form of consumer lighting available."*

  http://www.naturalnews.com/032020_eco-friendly_bulbs_toxic_chemicals.html#

  Is any of this common knowledge? The power of marketing / advertising is amazing - hearing only one side of the story.

- **Mold**
  Per the EPA, with the right conditions of moisture and oxygen, mold can grow on almost any organic surface. Mold is a toxin that can cause various health issues.
  http://www.epa.gov/mold/

  Keep in mind that there are non-toxic solutions for treating mold. See **'Other Toxic concerns'** heading at the end of Division 02 for some age old solutions.

  The mold remediation process commonly includes proper cleaning of items and proper removal and disposal of items that cannot be cleaned.

- **Radiation Exposure**
  Exposure to **Electromagnetic Fields (EMF)** and **Radio Frequency (RF)** radiation has been a concern especially as it relates to long term exposure to any electrical device or waves created from electronic equipment and wiring inside and the power lines outside buildings. Unfortunately cancer studies have also found this includes exposure to cell phone, Wi-Fi and other commonly used electronic devices. EMF and RF exposure in buildings appears to be causing the *next* Sick Building Syndrome.

The following websites are just a few related resources focusing on the concerns of EMF and RF:

- Electric & Magnetic Fields Radiation from Power Lines
http://www.epa.gov/radtown/power-lines.html

Notice however, that this link now redirects you to another EPA site seconds later. Why?

- *Urgent Warning to All Cell Phone Users*
http://articles.mercola.com/sites/articles/archive/2012/06/16/emf-safety-tips.aspx
includes video 1:01:40

Like with the warning assigned to a pack of cigarettes, it is obvious that the above cell phone concern has not been impacted many. As a society, people do as they will, not wanting to believe the warnings. What upsets me the most however is when a infant is teething on one, or a small child is given one to occupy his/her time. They don't know better, but the parents should . . .

- *Is Dirty Electricity Making You Sick?*
  http://www.prevention.com/health/healthy-living/electromagnetic-fields-and-your-health

- **Radon**

  Per the EPA, Radon is a naturally occurring odorless radioactive gas that can cause Lung Cancer. Radon Remediation is possible for existing construction and to address this is typically a code requirement for new construction.
  http://www.epa.gov/radon/

Radon Resistant New Construction
http://www.epa.gov/radon/rrnc/index.html

The remediation process for Radon commonly includes passive venting or in-line fan assisted venting from under a concrete slab for new and existing construction. See the International Code Council (ICC) for related information.

*- See Section **13 49 00 - Radon Protection,** for related products and information.*

- **Termites**

  Per the EPA, *"termites cause billions of dollars in structural damage".* To avoid further damage and to ensure complete and proper application, treatment should be by a State Licensed professional termite extermination service.
  https://www.epa.gov/safepestcontrol/termites-how-identify-and-control-them

Along with the EPA, additional information about **pesticides** is found at the:

National Pesticide Information Center
Phone: 1-800-858-7338
http://npic.orst.edu/

*See CSI Section **10 81 00 - Pest Control Devices** for related information*

Wood, Concrete *(and other)* **Restoration**

If a structural item needs restoration or replacement, a Structural Engineer should be involved.

Contact the below companies for a complete listing of wood, concrete, masonry and weather stripping related products and related videos.

- **Abatron, Inc.**
  5501 - 95th Avenue
  Kenosha, WI 53144
  Tel: (262) 653.2000
  ORDERS: 800.445.1754
  Fax: (262) 653.2019
  http://www.abatron.com/buildingandrestorationproducts.html

- **Advanced Repair Technology, Inc.**
  PO Box 510
  Cherry Valley, NY 13320
  Phone: 866-859-2787
  Fax: 215-348-8422
  http://www.advancedrepair.com/index.html

With the *Advanced Repair Technology*, the following link includes a *This Old House* video, using an Advanced Repair Technology wood restoration product.

http://www.thisoldhouse.com/toh/video/0,,20460611,00.html

- **PC Products**
  Protective Coating Co. *(Since 1954)*
  Allentown, PA 18102
  Phone: 800-220-2103
  Fax: 610-432-5043
  http://www.pcepoxy.com/

- **Additional concerns include:**

With **Facility Remediation**, cleaning products are a critical importance to regular maintenance. With the choice of such products, it is important to know that there are various **natural non-toxic remedies** available, including, but not limited to the age old solutions that use: ***Vinegar, Lemon Juice,*** and ***Baking Soda***. Because these are natural, they will not pollute the **Indoor Air Quality (IAQ)** and will eliminate the scare of having such products accessible to those (younger or older) that do not know better. See: **http://articles.mercola.com/sites/articles/archive/2014/03/31/spring-cleaning-tips.aspx**

As a final thought for **Division 02 - Existing Conditions**, it should be understood that everything needs some form of maintenance, so, for every building and related site, we are responsible to be good stewards of it for the time that we own, lease, and/or occupy it - until it is entrusted to someone else.

End of Division 02

Space for notes:

DIVISION 03 - Concrete

Concrete is one of the most widely tested and utilized materials in construction, be it for structural or decorative purposes. Because it is made and transported locally, and it is so versatile. With added steel reinforcement, it then can offer both compressive and tensile qualities often needed for site cast or precast applications which have made it a valued, long lasting, fireproof, and storm proof building material - that stands the test of time.

Exposed concrete surfaces can include exposed aggregate, or be colored, stamped, and/or polished.

A Structural Engineer should be involved to properly design the concrete foundation, floor, wall or roof system(s) and/or verify the structural strength of existing systems. Before doing so, it would be important to review International Building Code (IBC) **Chapter 18** - *Soils and Foundations,* and/or International Residential Code (IRC) **Chapter 4** - *Foundations*.

Use this website for *PremiumACCESS or* free *PublicACCESS* viewing of ***I-Codes*** *and* ***ICC Standards***:
https://codes.iccsafe.org/

03 00 00 - Concrete

According to, ***Specify Concrete.org***, concrete:

- *Contributes to LEED credits*
- *Is a wise choice for naturally reflective pavement, that does not require re-sealing*
- *Can be installed as a pervious pavement*
- *Is used to create durable buildings, using CMU, tilt-up, and/or pour-in-place applications with various finishes*
- *Can be used for decorative / functional interior or exterior applications such as stamped concrete flatwork, color stained and sealed concrete floors, countertops, and other imaginative applications.*

- **Specify Concrete.org** is a service of Pennsylvania Aggregate and Concrete Association (PACA).
http://www.specifyconcrete.org/

- **Pennsylvania Aggregate and Concrete Association** (PACA)
  http://www.pacaweb.org/

- **American Concrete Institute** (ACI)
  38800 Country Club Drive
  Farmington Hills, MI 48331
  Phone: 248-848-3700
  Fax: 248-848-3701
  http://www.concrete.org/ContactUs.aspx

03 05 00 – Integral Waterproofing of Concrete

- **Sika Corporation**
  201 Polito Ave.
  Lyndhurst, NJ 07071
  Phone: 800-933-SIKA

  https://usa.sika.com/en/concrete-products/concrete_home/concrete-admixture/watertight-concrete.html

A concrete admixture is one that prevents water from getting in or out by reducing water to cement ratio that increases ease of concrete placement while also reduces concrete porosity, thus reducing the possible ingress of water through the concrete.

03 05 13 – Coal Fly Ash

Fly Ash is one of the waste products of the coal-burning electric generating power plants. It is removed from the exhaust flues with the help of the required emissions-control scrubber equipment. However, as stated in *The Washington Post* on June 2, 2014, *"the Environmental Protection Agency proposed a regulation Monday that would cut carbon dioxide emissions from existing coal plants by up to 30 percent by 2030 compared with 2005 levels, taking aim at one of the nation's leading sources of greenhouse gases".*

It is understood that such reduction will reduce Fly Ash byproduct quantities as used in concrete. The substitution of Fly Ash allows the reduction of Portland Cement to manufacture various concrete and gypsum board products.

Per the **Indiana Decorative Concrete Network**:
http://indecorativeconcrete.com/?page_id=1445

- *Fly ash in concrete is considered to be a green product, as it reduces what would go into the landfills.*
- *Fly ash makes the concrete more workable as it lowers the Heat of Hydration, creating a longer set time.*
- *Concrete made with fly ash requires less water.*
- *Concrete made with fly ash is less permeable.*
- *Fly ash is typically less expensive than the straight cement mixes.*
- *Fly ash makes concrete stronger than common cement based mixes.*

*(See CSI Section **04 05 13** for additional information.)*

03 10 00 – Concrete Forming and Accessories

- **Atlas Construction Supply**
  4640 Brinell Street
  San Diego, CA 92111
  Phone: 858.277.2100
  Fax: 858.277.0585
  http://www.atlasform.com/

With other locations in: Denver, CO; Las Vegas, NV; Phoenix, AZ; Fullerton, CA; and Denver, CO

Atlas Construction Supply offers a variety of concrete forming and curing agents including but not limited to:

- Atlas Bio Guard™, a **100% biodegradable, zero-VOC vegetable oil** concrete form release agent.

- Atlas Quantum-Cure™, a clear, odorless, **zero-VOC, water based, environmentally safe** product that is used to help retain moisture control during the concrete curing process.

- **Caraustar**
  Industrial & Consumer Products Group
  3100 Joe Jerkins Boulevard
  Austell, GA, 30106
  Phone: 770-948-3101
  Email: constructionproducts@caraustar.com
  http://www.caraustar.com/

*"**Caraustar** is a leading producer of fiber tubes used to form concrete columns, footings, piers and other structures. Caraustar **KlimatePRO™easy-POUR™** forms are engineered to provide outstanding performance under tough jobsite conditions and even in inclement weather. Produced using 100% Recycled Paperboard, our multiple manufacturing facilities located across the United States and Canada put us close to where you need us most."*

- **Dayton Superior**

Corporate Headquarters
1125 Byers Road
Miamisburg, OH 45342
Toll Free: 800.745.3700
Phone: 937.866.0711

http://www.daytonsuperior.com/

*"**Dayton Superior** believes it is time for our industry to renew and expand our dedication to American manufacturing. With 98% (and growing) of our products Made In America, Dayton Superior is dedicated to creating and retaining jobs in the United States. You can rely on the quality of our products and services...manufactured and provided in America, by Americans,* (including) *the **Symons**® brand of forming systems has stood as the preferred system for concrete forming around the globe."*

- **Form-A-Drain** CertainTeed Corporation
  P.O. Box 860
  Valley Forge, PA 19482
  Professional: 800-233-8990
  Consumer: 800-782-8777

http://www.certainteed.com/

Certainteed Corp. manufactures **Form-A-Drain.** A **stay-in-place concrete formwork** that is able to provide **perimeter drainage** and also **radon mitigation, using one pipe**.

See:
http://www.certainteed.com/resources/FAD%20General%20Brochure%2040-95-07J.pdf

*- See CSI Section **13 49 00** - **Radon Protection**, for related information*

03 11 19 – Insulating Concrete Forming

There are mixed reviews to Insulating Concrete Forms (ICFs). It probably all comes down to how each manufacturer designs their block system, and potentially the number of stories involved. I personally witnessed a 5-Story hotel built a few years ago using ICFs. The manufacturer stated that after the foundation, each floor could be built in a week. Well, not only did that not happen, but because of concrete blow-outs and extreme bulges, one of the sub-contractors stated that it took them several months to rasp the walls so that they were sufficiently straight. The hotel was completed several months after the original due date.

Even with the above experience, as wall, floor and roof forming product, ICFs seem to be the way to go - to achieve a substantial R-value and for the strength of concrete to stand up against disaster. Here are just a few of the manufacturers representing various regions in the USA. Most use Expanded Polystyrene (EPS), but wood chip-cement is also an option. As always, it would be wise to check references and see past projects that use the product that is being considered.

- **Faswall**

  Shelter Works, Ltd.
  P.O. Box 1311
  Philomath, OR 97370
  Phone: 855.558.4588

  http://faswall.com/

  *"With a 25-year track record, **Faswall®** insulated wood chip-cement wall forms are for experts and do-it-yourself builders alike. The organic building materials are simpler, more sustainable, longer-lasting, more affordable, safer, and healthier than traditional materials, . . . (and) no polystyrene plastics or foams."*

  Note: Similar to Durisol, however Faswall® is made in the USA.

- **Lite Form Technologies**

  1950 W.29th St.
  South Sioux City, NE. 68776
  Phone: 402-241-4402
  Phone: 800-551-3313
  Fax: 402-241-4435

  http://www.liteform.com/

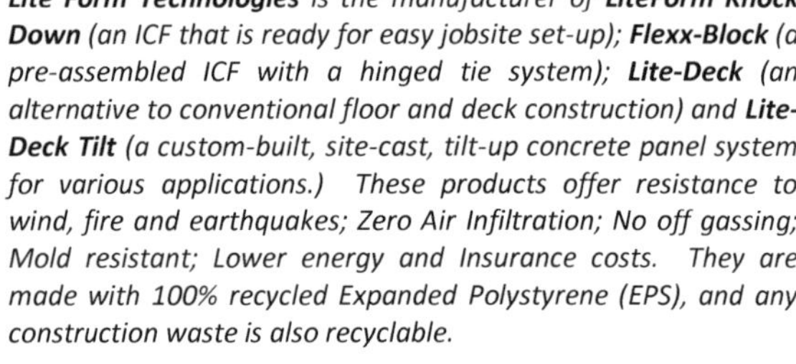

  Lite Form Technologies *is the manufacturer of* ***LiteForm Knock Down*** *(an ICF that is ready for easy jobsite set-up);* ***Flexx-Block*** *(a pre-assembled ICF with a hinged tie system);* ***Lite-Deck*** *(an alternative to conventional floor and deck construction) and* ***Lite-Deck Tilt*** *(a custom-built, site-cast, tilt-up concrete panel system for various applications.) These products offer resistance to wind, fire and earthquakes; Zero Air Infiltration; No off gassing; Mold resistant; Lower energy and Insurance costs. They are made with 100% recycled Expanded Polystyrene (EPS), and any construction waste is also recyclable.*

- **SmartBlock™**

  ConForm Global
  PO Box 2839
  Coeur d'Alene, ID 83816
  Phone: 1-800-CONFORM (1-800-266-3676)
  Fax: 1-208-676-0615
  E-mail: info@smartblock.com

  http://www.smartblock.com/index.html

  "ConForm Global produces a variety of ***SmartBlock™*** *Insulating Concrete Form heights and various ties to create multiple widths.* ***SmartBlock™*** *EPS forms can contribute more than 25 points towards the 69 available points for a LEED NC certified structure".*

It should be noted that Expanded Polystyrene (EPS) should be treated to prevent termite problems. Contact:

- **EPS Industry Alliance**
  1298 Cronson Boulevard
  Suite 201
  Crofton, MD 21114
  Phone: 800.607.3772
  Fax: 410.451.8343
  http://www.epsindustry.org/building-construction/insect-infestation

- **Thermomass**
  1000 Technology Drive
  Boone, Iowa 50036
  Toll Free: 800.232.1748
  Phone: 515.433.6075
  Fax: 515.433.6088

  http://www.thermomass.com/

  Thermomass provides various insulated tilt-up pre-cast and site-cast insulated concrete wall systems for residential homes on up to stadium construction. They are hurricane, tornado, fire resistant, quiet and almost maintenance free.

03 20 00 – Concrete Reinforcing

- **Byer Steel**
  200 W. North Bend Road
  Cincinnati, Ohio 45216
  Phone: 513-821-6400, or 800.543.4922
  Fax: 513-821-6915

  http://www.byersteelminded.com/

  Byer Steel offers a variety of steel products, including rebar and welded wire fabric made from the recycled steel service the company operates.

- **Helix Steel**
  Corporate Office
  300 N. Fifth Avenue, Suite 130
  Ann Arbor, MI 48104
  Phone: 734-322-2114
  Fax: 734-786-1633

  http://www.helixsteel.com/

  *"**Helix**, simply put, is a **replacement for rebar**" . . . and is "the only discontinuous concrete reinforcement product in the world that now has a ISO certified design manual that can be followed to design vertical applications (such as walls) with Helix as the primary concrete reinforcement".* Helix offers increased durability, shear and crack resistance to concrete - and **reduced labor costs**.

- **Kodiak Fiberglass Rebar**
  6500 Hwy 321
  Dayton, TX 77575
  Phone: 936-257-1800
  Fax: 936-257-8800

  http://www.kodiakrebar.com/

Kodiak Rebar is 100% made in USA and contain 100% A-Grade American Made Materials. **Fiberglass Rebar** is a substitute for the typical steel rebar.

Unfortunately, **steel corrodes, and then it expands**, causing spawling of the concrete, which then ultimately causes severe damage, and potential failure through time.

*It should be noted that a potential alternative to fiberglass rebar is **Hot-Dip Galvanized rebar** or **Epoxy-coated rebar**.*

See the below website for more information:
https://www.galvanizeit.org/uploads/publications/Galvanized_Rebar_vs_Epoxy_Rebar.pdf

- **American Galvanizers Association** (AGA)
  6881 South Holly Circle, Suite 108
  Centennial, Colorado 80112
  Phone: 720-554-0900
  Fax: 720-554-0909

  www.galvanizeit.org

  "The AGA is a non-profit trade association dedicated to serving the needs of after-fabrication galvanizers, fabricators, architects, specifiers, and engineers. The AGA provides technical support on today's innovative applications and state-of-the-art technological developments in hot-dip galvanizing for corrosion control."

- **Nucor Corporation**
  1915 Rexford Road
  Charlotte, NC 28211

  http://www.nucorbar.com/

  Nucor Corporation has some of the most efficient mills in the USA that uses recycled steel and other metals to manufacture **steel rebar,** along with other **extruded steel products** at multiple locations including:
  Auburn, NY; Birmingham, AL; Jackson, MS; Bourbonnais, IL; Memphis, TN; Jewett, TX; Norfolk, NE; Plymouth, UT; Seattle, WA; Kingman, AZ; Wallingford, CT; Hammond, IN; Marion, OH; Darlington, SC

- **Wire Products of Florida**
  Corporate Office/ Plant Operations
  4300 Northwest 10th Avenue
  Fort Lauderdale, FL 33309
  Phone: 800-457-1018

  https://www.wireproducts.us/

  Wire Products of Florida produce various American made steel products, including **reinforcing bars**, tools and **steel wire**.

*- See **Division 05 – Metals,** for related information.*

03 30 00 – Cast-in-Place Concrete

Contact local suppliers and finishers, and/or:

- **National Ready Mixed Concrete Association**
  Education/Technical Resources/Sustainability
  Lionel Lemay, P.E., SE, LEED AP
  Senior Vice President, Sustainable Development
  1244 Crane Boulevard
  Libertyville, IL 60048
  Phone: 847-918-7101

  http://www.nrmca.org/about/staff_bios.asp

  *"Founded in 1930, the **National Ready Mixed Concrete Association** is the leading industry advocate. Our mission is to provide exceptional value for our members by responsibly representing and serving the entire ready mixed concrete industry through leadership, promotion, education and partnering to ensure ready mixed concrete is the building material of choice."*

- **Symons** *(owned by Dayton Superior)*
  Dayton Superior
  Corporate Headquarters
  1125 Byers Road
  Miamisburg, OH 45342
  Phone: 800-745-3700 937-866-0711
  http://www.daytonsuperior.com/forming.aspx

  *The reliable **Symons** products provide a wide variety of **reusable formwork** and **form liners** - for various cast in place applications.*

  See the various patterns & textures details, and end results after using **concrete form liners**.

  http://www.daytonsuperior.com/products/system?name=formliner-system

- **UFP Concrete Forming Systems**
  UFP Parker, LLC
  116 N. River Ave.
  Parker, PA 16049
  Phone: 724 321-3688

  http://www.ufpconcrete.com/en/Products-and-Services/Plywood.aspx

  For all locations across the USA, see:
  http://www.ufpconcrete.com/en/Our-Locations.aspx

  ***UFP Concrete Forming Systems** provides a wide variety of **reusable wood panel products** used in concrete formwork.*

03 40 00 – Precast Concrete

- **National Precast Concrete Association**
  1320 City Center Drive, Suite 200
  Carmel, IN 46032

Phone: (800) 366-7731
Fax: (317) 571-0041

http://precast.org/

NPCA provides the latest news in precast concrete, including an informative website; webinars, coursework; concrete plant certification; the various products available within the industry, two leading magazine/journals, and other industry resources.

- ## Spancrete Manufacturers Assoc. (SMA)

  8989 N. Port Washington Road
  Milwaukee, WI 53217
  Phone: 414-351-5588
  Fax: 414-351-4617

  http://www.spancrete-sma.com/applications.php

  "SMA member manufacturers take seriously the design, engineering, production and construction of some of the best building materials available today. All operate under the rigid manufacturing guideline set for the production of ***Spancrete® hollowcore.*** *All (***Spancrete pre-stressed planks*** *are) made using a slipforming process with machinery developed, patented and manufactured by the Spancrete Machinery Corporation. A Better Way to Build."*

- ## Superior Walls Precast Walls

  937 East Earl Road
  New Holland, PA 17557

  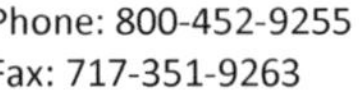

  Phone: 800-452-9255
  Fax: 717-351-9263

  http://www.superiorwalls.com/

  Precast concrete **below grade** and **above grade** systems offer:
  - Reduced energy leakage
  - Lowers energy costs
  - Saves valuable resources
  - Improves indoor air quality
  - No on-site soil contamination (from form release agents)
  - No on-site damp proofing
  - Reduces jobsite waste
  -National Green Building Standard Certified
  - Energy Star
  - USGBC Member

03 41 10 – Precast Autoclaved Aerated Concrete

Simply known as **Autoclaved Aerated Concrete** (AAC), or sometimes known as **Autoclaved Cellular Concrete** (ACC), this durable, load-bearing, naturally fire retardant & insulating, light-weight material can be understood to be a *unit masonry product*, so it has also been found to be assigned to the ***04 20 00 - Unit Masonry*** section. However, if placed there, it gets lost amongst other products. Deciding to use this designation provides a separate, stand-alone heading.

- **Aercon Autoclaved Aerated Concrete**
  Aercon AAC
  3701 CR. 544 East
  Haines City, Florida 33844
  Phone: 800.830.3171
  Fax: 863.422.6361

http://www.aerconaac.com/

Aercon AAC products and finished projects offer:
- Fire Resistance / Non-combustible construction
- Superior Thermal Insulation
- Excellent Acoustical Performance
- Durable - without the use of Fly Ash
- Design Versatility & Flexibility
- Reduced Time & Labor costs
- Environmentally Friendly
- 30% lighter, and 50% larger than standard CMU

Autoclaved Aerated Concrete (Sometimes known as *Air Crete*) -
Discovery Channel video.:
https://www.youtube.com/watch?v=empEMs-2zmA 6:57

- **International Masonry Institute** (IMI)
  42 East Street
  Annapolis, MD 21401
  Phone: 800-IMI-0988

http://imiweb.org/

IMI has valuable information on their website, including a ***Masonry Detailing Series***, illustrating common masonry wall systems drawn.

See this document for valuable content regarding ACA: **http://imiweb.org/wp-content/uploads/2015/10/01.02-AAC-MASONRY-UNITS.pdf**

- **Portland Cement Association** (PCA)
  5420 Old Orchard Road
  Skokie, Illinois 60077-1083
  Toll Free: 847.966.6200
  Fax: 847.966.9781

http://www.cement.org/think-harder-concrete-/homes/building-systems/autoclaved-aerated-concrete

03 50 00 – Cast Decks and Underlayment

- **Arcis Corp**
  Headquarters / East Coast Manufacturing Facility
  Phone: (503) 647-5042
  Fax: (503) 647-7731

http://www.arcis-corp.com/

Phyisical Address:
10680 NW 289th place
North Plains OR 97133

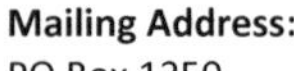

Mailing Address:
PO Box 1250
North Plains OR 97133

Arcis Corp. offers cost savings over wood, composite, conventional concrete and includes **Thin Pre-Stressed Cladding**; **Thin Pre-Stressed Decking**; **Detectable Warning Panels** (for sidewalks); and **Custom Precast Concrete.** Arcis can distribute decking throughout the US and Canada.

- ## Level-Right

Maxxon® Corporation
920 Hamel Road
P.O. Box 253
Hamel, MN 55340
Phone: 800.356.7887 or (763) 478-9600
Fax: (763) 478-2431

http://www.maxxon.com/level-right/data

Level-Right®self-leveling, thin-topping Cementitious underlayment from Maxxon® ***(formerly the Gyp-Crete Corp.)*** is not polymer-modified, resulting in **nearly zero VOC's** and also contains **over 51% fly ash**. It can be featheredged and is GreenGuard Gold Certified.

- ## Lite Deck

Multiple USA locations
1-800-551-3313

http://www.litedeck.com/

The **LiteDeck®** ICF System is a lightweight, stay-in-place form made of Expanded Polystyrene (EPS) and used to construct **site-cast or precast concrete floors, roofs, decks, and walls** for **commercial, industrial and residential uses.** Lite-Deck® integrates EPS Insulated Forms and uses reinforced structural concrete that stand up to extreme storms.

The website provides **Detailed Drawings** and **related videos.**

- ## Molin Concrete Products

415 Lilac St.
Lino Lakes, MN 55014
Phone: 651-786-7722
Phone: 1-800-336-6546
Fax: 651-786-0229

http://www.molin.com/

Molin Concrete Products provides precast concrete applications for Single Family, Multi-Family, Hotel/Motel, Manufacturing, Healthcare, Offices, Public / Governmental Buildings, Recreational / Casino, & Automotive Services

03 60 00 – Grouting

For a selection of **03 60 00** related LEED products, see:

http://www.arcat.com/divs/sec/sec036000.shtml

Concrete: *Some additional information*

- **Preservation of Historic Concrete**
  National Park Service
  US Department of the Interior
  Technical Preservation Services
  Preservation Brief 15

  https://www.nps.gov/tps/how-to-preserve/briefs.htm

End of Division 03

Space for notes:

DIVISION 04 - Masonry

Like Concrete, masonry products (typically **brick, concrete block, stone** and **glass block**) along with their accessories, are widely tested and utilized materials for structural and/or aesthetic applications in the AEC industry. Because masonry products are typically quarried, manufactured, and transported locally, while being versatile, using various masonry bond patterns and/or the added steel reinforcement (commonly found horizontally and/or vertically within the mortar joints) and/or using masonry wall tie applications (to attach a masonry façade to the structure), has made it a valued, long lasting, fire proof building material. It has only been in the most recent years that artificial stone (made with concrete) or actual thin stone veneer (cut from actual stone) has been manufactured for "stick on" cosmetic applications. This thin stone veneer requires no reinforcing other than sometimes chicken wire, thus less labor as compared to traditional 4" masonry veneer.

04 01 00 - Maintenance of Masonry

Everything needs some maintenance, even masonry. One of the common needs is **tuck pointing**. This is the term used to remove and replace the mortar between the brick, block or stone. But, this is not one of those projects that is needed on a regular basis. For instance, a chimney might need such after 50yrs, and it may not. There are various factors to consider, including the type of mortar that was used. Depending on the age of the structure, the mortar used prior to 1900 AD - could be lime based vs. mortar that has Portland cement added. *See CSI Section **04 05 13 - Masonry Mortaring,** for this related topic.*

I can remember years ago, when sand blasting was common to remove paint off of brick buildings. Well, sand removed the paint and grime, but unfortunately it permanently damaged the brick as well. See the following *Preservation Briefs* for proper techniques.

- **Cleaning and Water-Repellent Treatments for Historic Buildings**
  National Park Service
  US Department of the Interior
  Technical Preservation Services
  Preservation Brief 1
  https://www.nps.gov/tps/how-to-preserve/briefs.htm

- **Dangers of Abrasive Cleaning To Historic Buildings**
  National Park Service
  US Department of the Interior
  Technical Preservation Services
  Preservation Brief 6
  https://www.nps.gov/tps/how-to-preserve/briefs.htm

- **Removing Graffiti from Historic Masonry**
  National Park Service
  US Department of the Interior
  Technical Preservation Services
  Preservation Brief 38
  https://www.nps.gov/tps/how-to-preserve/briefs.htm

04 05 13 - Masonry Mortaring

With this heading section, depending on its application, there is a choice between **Portland cement** or **Lime based mortars** – along with the

opportunity to substitute **Fly Ash** to reduce cement content.

Natural hydraulic **Lime based mortar** was commonly used in historic structures and today, typically should be used in *"performing professional **masonry restoration** and to build **sustainable structures**"* of the future - as described at the Lime Works website below.

Basically, the only companies that distribute natural hydraulic Lime based mortar in the US are:

- **Lime Works** (Eastern USA)
  PO Box 151
  Milford Square, PA 18935-0151
  Phone: 215-536-6706
  Fax: 215-536-2281

  http://www.limeworks.us/home.php (with video)

  Local retailers of the above are found at:
  http://limeworks.us/retailers.php

- **TransMineral USA** (Central USA)
  Denver, CO
  707-769-0661
  Email TransMineral USA
  http://www.limes.us/
  http://www.limes.us/distributors/

- **TransMineral USA** (Western USA)
  Oakland, CA
  707-769-0661
  Email TransMineral USA
  http://www.limes.us/
  http://www.limes.us/distributors/

The **Lime based mortar** from the above companies has been, and continue to be used to build new and restore many US structures, such as: St. Patrick's Cathedral, NYC; Rotunda at University of Virginia, Charlottesville, VA; and others shown at: **http://www.limes.us/projects/** (with videos)

Mortar: *Some additional information*

- **Repointing Mortar Joints In Historic Masonry Buidlings**
  National Park Service
  US Department of the Interior
  Technical Preservation Services
  Preservation Brief 2

  https://www.nps.gov/tps/how-to-preserve/briefs.htm

Fly Ash, a by-product of coal-fired electric generating plants and steel manufacturing has been used to reduce Portland cement content in concrete and concrete masonry products. *(See CSI Section **03 05 13** for additional information.)*

A related **Fly Ash based article** *(Is It A Danger Or Not??)* from *Engineering News-Record*, can be found at:
http://enr.construction.com/opinions/editorials/2010/04 14-FlyAshDilemma.asp

04 05 23 - Masonry Accessories

- **Archovations, Inc.**
  701 Second Street
  Hudson, WI 54016
  Phone: 715.381.5773
  Fax: 715.381.9883
  http://archovations.com/masonry-mat/

  CavClear® Masonry Mat is made from 100% recycled plastic and once installed full-height behind brick, it eliminates weep obstructions for a continuous drainage path to prevent moisture

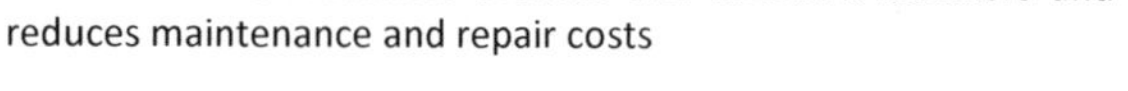

intrusion while minimizes thermal and moisture transfers and reduces maintenance and repair costs

- **Conproco Masonry & Concrete Repair and Preservation Products**
  Conproco
  17 Production Drive
  Dover, NH 03820

  LEED Credit information:
  Trish Veneziano,
  Phone: 800.258.3500, Ext. 541
  email: tveneziano@conproco.com

  http://www.conproco.com/

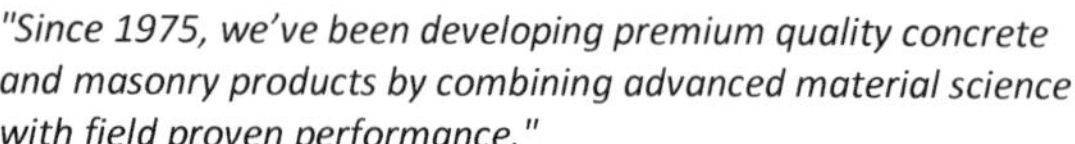

"Since 1975, we've been developing premium quality concrete and masonry products by combining advanced material science with field proven performance."

Per there website, **Conproco** enables: Concrete Repair and Preservation; Masonry Repair and Preservation; Protective Coatings; and Exterior Wall Systems.

- **Hohmann & Barnard, Inc.**
  800-645-0616
  Locations in Alabama, Illinois, Maryland, New York, Utah, Texas
  http://www.h-b.com/

Hohmann & Barnard, Inc. provides over 79 years of proven quality and technical expertise to the masonry industry that includes but not limited to Masonry Anchors & Ties; Lintels; Expansion Joints / Control Joints; Reinforcement; Air & Vapor Barriers; Weep Holes & Vents; Masonry Restoration & Repair Systems; Termite Shields; Reglets & Termination Bars.

- **Mortar Net Solutions**

326 Melton Rd.
Burns Harbor, IN 46304
Phone: 800.664.6638
Fax: 219.787.5088
http://www.mortarnet.com/

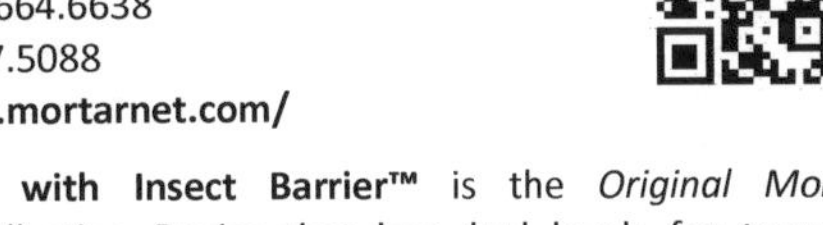

MortarNet® with Insect Barrier™ is the *Original Mortar Dropping Collection Device* that has dual levels for trapping mortar droppings that would otherwise prevent any moisture from properly dropping to the weep holes. Helps prevent efflorescence and damage from freeze-thaw cycles, and the recycled content helps with LEED Certification.

- **South Atlantic Masonry Products**

4162 South Creek Road
Chattanooga TN 37406
Phone: (877) 721-3355
Fax: (423) 826-3524
http://masonry.southatlanticllc.com/

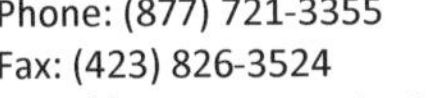

South Atlantic Masonry Products produces **Hot Dipped Galvanized products, including Steel Lintels; Wall Ties; Anchor Bolts; Metal Lath, and other related hardware.**

*- See Section **03 20 00 - Concrete Reinforcing** and **Division 05 – Metals**, for related information.*

04 20 00 - Unit Masonry

Masonry Units commonly are understood to be **modular brick**, **block** (CMU), and **glass block** systems. Meaning, these are manufactured in standard modular sizes, much like Lego® products. These sizes are commonly based on nominal 8" (ie: .667') & 16" (ie: 1-4" or 1.333') modules, yet, there are exceptions. If designing with such, it is wise to layout the structure using the material and its module, (and not fight it), so to maximize the material and not waste time and money cutting it to fit. (Would you do such with a Lego® set?) Commonly, various Division 08 - Openings (windows, doors, etc.) are manufactured to fit such masonry opening (M.O.) sizes. By understanding and working with the material, the results offer little construction debris. In fact, it can offer a larger building for less cost. For instance, a rectangular building could be 33'-0" x 62'-6" (2062.5 sq ft). But, if the building was resized to be 33'-4" x 62'-8" (2088.889 sq ft), using a masonry module in plan for all measurements, including the masonry openings, the building is larger, offers no masonry cutting, so potentially is taking less time, and potentially less cost (including less debris to the landfill) . . . as a result of such planning. (ie. Sustainable Design!)

And, it should be said that if building offsets are desired, verses a standard rectangular building, this can be done as well. Just keep the below in mind with <u>all</u> dimensions:

- *If the feet are <u>even</u> (2', 12', 126'), then the inches should be either 0" or 8". (ie. 2'-0" or 2'-8"; 12'-0" or 12'-8"; 126'-0" or 126'-8")*
- *If the feet are <u>odd</u> (1', 13', 127'), then the inches should only (and always) be 4". (ie. 1'-4"; 13'-4"; 127'-4")*
- *The above applies to vertical and horizontal dimensions.*
- *See this site (and others) for typical <u>actual</u> block sizes:* ***http://www.alliedconcrete.com/wp-content/uploads/2010/04/CMU-Catalog.pdf***
- *See this site (and others) for typical <u>actual</u> brick sizes:* **http://www.beldenbrick.com/brick-dimensions-guide.asp**

- With the actual sizes of block and brick, mortar joints are many times 3/8" wide to maintain the 8" or 16" nominal unit size.

Now that you understand this concept, have fun. There is a nice variety of textures, patterns, and colors to choose from, including being creative with various **masonry bond patterns**, shown at this website: **http://www.gobrick.com/portals/25/docs/technical%20notes/tn30.pdf**

The following are a sampling of some unit masonry manufacturers, listed after *The Brick Industry Assoc.* & the *National Concrete Masonry Association.*

- (The) **Brick Industry Association**
  1850 Centennial Drive, Ste. 301
  Reston, VA 20191
  Phone: 703-620-0010

  http://www.gobrick.com/

The Brick Industry Association is the *"**National Authority of Clay Brick Construction**"*. It produces well detailed publications that are technical and inspirational to the industry. One was provided above, discussing masonry bond patterns. Another, shown below, communicates typical details for cavity walls. See: **https://www.gobrick.com/portals/25/docs/technical%20notes/tn21b.pdf**

- ## National Concrete Masonry Association

13750 Sunrise Valley Drive
Herndon, VA 20171
Phone: 703-713-1900
Fax: 703-713-1910
Email: info@ncma.org

http://www.cement.org/learn/materials-applications/masonry/masonry-construction

"NCMA is the national trade association representing the producers and suppliers of concrete masonry products, including concrete block, manufactured stone veneer, segmental retaining walls and articulating concrete block."

- ## (The) Beldon Brick Company

PO Box 20910
Canton, Ohio 44701-0910
Phone: 330-451-2031

http://www.beldenbrick.com/about/brick-company-sustainability

*"**The Belden Brick Company** traces its roots to the Diebold Fire Brick Company organized in Canton, Ohio in 1885 by Henry S. Belden and four associates. In the US brick industry, The Belden Brick Company is the sixth largest (by production volume) manufacturer."*

According to its website, it manufactures: **Face Brick; Thin Brick; Oversized Brick; Structural Brick; Pavers**; and **Special Shapes**.

- ## Integra Wall Systems

4150 West Turney Street
Phoenix, AZ 85019
Phone: 602.352.3500
Toll Free: 800.366.7877

http://www.integrawall.com/index.htm

The **Integra Wall System** is a structural post-tensioned concrete masonry wall system that applies to various residential and non-residential single-wythe applications allowing for an economical, durable, and energy saving wall construction that incorporates a unique internal threaded rods and closed cell foam insulation, while using an H-style CMU block.

- ## Omni Block & Omni Brick

5940 S. Rainbow Blvd. , Suite 1003
Las Vegas, NV 89118
Phone: 877.711.6664

http://www.omniblock.com/ with 10:17 video

The **Omni Block** and **Omni Brick** conserves raw materials while providing energy savings using engineered foam inserts that minimize thermal conductivity through the block. Available in many States.

- **Trenwyth** (An Oldcastle Company)
  Headquarters &
  Manufacturing Facility:
  Trenwyth Industries
  One Connelly Road
  Emigsville, PA 17318
  Phone: 800.233.1924

  Manufacturing Facility:
  Trenwyth Industries
  4626 North 42nd Avenue
  Phoenix, AZ 85019
  Phone: 800.331.9823

  http://www.trenwyth.com/

Trenwyth products include:

- **Acousta-Wal** sound-absorbing masonry units
- **Astra-Glaze-SW+®** glazed concrete masonry units *(Illustrated)*
- **InsulTech™** Insulated Concrete Masonry System
- **Mesastone** textured masonry units
- **Mojavestone** masonry units
- **Monumental** over-sized masonry units
- **Split-Face** masonry units
- **Trendstone** ground face masonry units
- **Trendstone Plus** filled and polished ground face masonry units
- **Verastone** recycled ground face masonry units
- **Verastone Plus** recycled filled and polished ground face masonry units

NOTE: *If looking for* ***Autoclaved Aerated Concrete****, See CSI Section* ***03 41 10 - Precast Autoclaved Aerated Concrete***

04 21 29 - Terra Cotta Masonry

Terra Cotta, (which means *'cooked earth'*, in Italian), is slowly becoming popular again as its natural sustainable characteristics are being rediscovered. It is being replicated and used in building restoration, and integrated on new structures. Some valuable resources are:

- **The Preservation of Historic Glazed Architectural Terra Cotta**
  National Park Service
  US Department of the Interior
  Technical Preservation Services
  Preservation Brief 7

  https://www.nps.gov/tps/how-to-preserve/briefs.htm

- **Boston Valley Terra Cotta**
  6860 South Abbott Road
  Orchard Park, NY 14127

Toll Free 888.214.3655
Phone 716.649.7490
Fax 716.649.7688

http://bostonvalley.com/

*"**Boston Valley Terra Cotta** is the leading manufacturer of custom architectural terra cotta for restoration of historic facades and creation of high performance building envelopes".*

For additional **04 21 29** related LEED products, see:
http://www.arcat.com/divs/sec/sec042129.shtml

04 23 00 - Glass Unit Masonry

- ### Pittsburgh Corning Glass Block

  Pittsburgh Corning Corp.
  800 Presque Isle Drive
  Pittsburgh, PA 15239
  Phone: 724.327.6100

  http://www.pghcorning.com/products/glass-block/

  "Glass block, made largely from sand and limestone, is 100 percent recyclable, inert, low maintenance, and highly durable. Yet its dynamic relationship with light gives architects the opportunity to create both aesthetically pleasing and energy efficient spaces. As a proven, natural material, glass block can add beauty and inspiration to a project while playing a significant role in sustainable design."

- ### Technical Glass Products - Glass Ore

  TGP
  8107 Bracken Place SE
  Snoqualmie, WA 98065
  Toll Free: 800.426.0279
  Fax: 800.451.9857
  http://www.tgpamerica.com/decorative/glassore/

  TGP Handcast Glass Bricks offer the weight and texture of masonry, yet the **Glass Ore** *"dramatically changes the appearance of a room when lit from behind or underneath. Ideally suited for artistic accents, **Glass Ore** can be used in **interior or exterior** applications".*

*- See CSI Section **08 80 00 - Glazing**, for related products and information.*

04 24 00 - Adobe Unit Masonry

- ### The Preservation of HIstoric Adobe Buildings

  National Park Service
  US Department of the Interior
  Technical Preservation Services
  Preservation Brief 5

https://www.nps.gov/tps/how-to-preserve/briefs.htm

- ## Adobe Block Manufacturing

El Pueblito Estates, Inc.
4820 Araceli Avenue
El Paso, TX 79938
Phone: 915-383-0188

http://www.adobemfg.com/sys/cbp/cbp.asp

*"**El Pueblito Estates, Inc.** is the only company in El Paso, TX that manufactures top quality sun-dried Adobe Block available in traditional, semi-stabilized, and fully-stabilized varieties.*

We manufacture our own quality controlled Adobe Block! We can produce 4500 adobe blocks per day. Our standard Adobe Block Size is: 14" Long x 10" Wide x 4" High. We do not use any straw. We mix a weighted amount of sand, clay with water to make a perfected and standardized adobe block.

See one of their manufacturing videos at:
https://www.youtube.com/watch?v=1NqW9lFR9b4 5:28

- ## Arizona Adobe Co.

Cordes Junction, AZ
Tucson, AZ
Phone: 928-254-0764

http://arizonaadobe.com

*"**ARIZONA ADOBE COMPANY's** primary focus is to supply adobes to earthen based contractors and builders. We are committed to providing affordable, beautiful and sustainable adobes. As adobe* 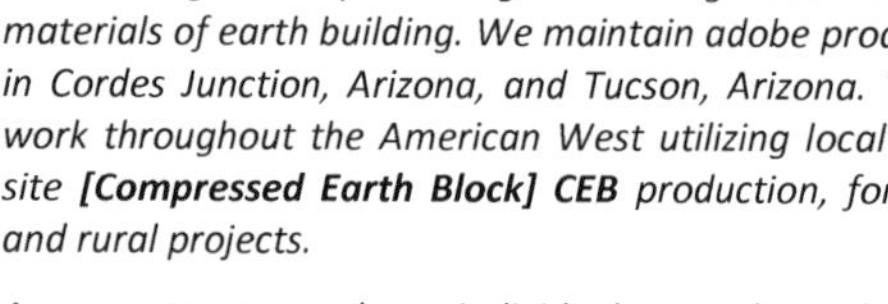 *brick makers, we take great pride and commitment to maintaining and preserving the indigenous methods and materials of earth building. We maintain adobe production yards in Cordes Junction, Arizona, and Tucson, Arizona. We regularly work throughout the American West utilizing local soils for on-site **[Compressed Earth Block] CEB** production, for both urban and rural projects.*

As a company and as individuals, we have long standing commitments to environmental and conservation stewardship along with unwavering support for local sustainable communities. We enjoy working with individuals and communities alike to develop, preserve and restore earth buildings."

Tucson Adobe video:
https://www.youtube.com/watch?v=tUhYCGEEUBQ 2:21

04 40 00 – Stone Assemblies

Over the past decade or so, (man-made) **cultured stone** veneer (also known as manufactured stone, or more specifically, Adhered Manufactured Stone Masonry Veneer) has become very popular. As a thin veneer, application is very easy. It tries to replace the long standing 4" veneer stone, that required a foundation, similar to a 4" brick veneer wall assembly. However, since it is concrete, there has been some concern to its durability over time - including breakage and fading. See what the warranty states. This will probably give a realistic understanding to what would be expected. Because of these concerns, cultured stone is not listed below. *See related article for more information:* **https://www.nachi.org/mfgd-stone-water-damage.htm**

- **Indiana Limestone Institute of America, Inc.**

  1502 I Street
  Suite 400
  Bedford, IN 47421
  Phone: 812.275.4426

  http://www.iliai.com/

  *"**The Indiana Limestone Institute** is a resource for architects, contractors, building owners and others seeking accurate, unbiased information about the use of Indiana Limestone in construction.*

  Indiana Limestone is sustainable, durable/long lasting, recyclable, offers low life-cycle costs, and uses an environmentally friendly production process that has been applied to *"countless government, commercial, educational and residential projects - including numerous U.S. state capitols and many of the government buildings in our nation's capitol."*

- **Real Stone**

  P.O. Box 637
  335 Park Ave., NW
  Bagley, MN 56621
  Fax: 218.694.6208
  Phone: 218.694.6207

  http://www.realstonesupply.com/

  Real Stone cuts natural Mississippi Headwaters Glacial Till granite for Commercial and Residential applications, including exterior building construction, outdoor living, landscape, interior flooring and other types of veneer finishes. This provides an option to the concrete stone veneer products that can fade. Applications can be for building, landscape and floors, and more.

- **Texas Quarries**

  Limestone & Stone
  P. O. Box 820
  Cedar Park, TX 78613
  Phone: 512.258.1474
  Fax: 512.258.0808

  http://www.texasquarries.com/

  Some of the Texas Quarries' products include thin cut Riviera Stone, Dimensional Limestone, Fast Track Stone ™ and custom limestone carving that provides a quality surface for residential, commercial, governmental, educational or institutional project.

  "With a customer list that includes clients from New York to California and from Canada to Japan, Texas Quarries provides superlative craftsmanship and unique limestone in a variety of styles and textures as expansive as the architect's imagination."

04 50 00 - Refractory Masonry

- **Alsey Refractories Company**
  Alsey Corporate Office
  1600 South Brentwood Blvd., Ste. 201
  St. Louis, MO 63144
  Phone: 314.963.7900

  Manufacturing Plant & Research
  266 State Route 106 South
  Alsey, Illinois 62610
  Phone: 314.963.7900

  http://alsey.com/products/category/residential/

  "From the makings of an authentic, custom pizza oven, to furnaces , from the most demanding industrial applications, everyone of our products bear signature of unparallel quality, pride and strength. Whether you're interested in private labeling or distribution, a company with needs for industrial refraction use, or searching for the best refractory products in residential building, look no further than the first name in firebrick - ***Alsey Refractories Company****."*

04 60 00 - Corrosion-Resistant Masonry

- **Belden Brick Company**

  P.O. Box 20910
  Canton, Ohio
  44701-0910
  Phone: 330.451.2031

  http://www.beldenbrick.com/

  "Belden shale and fire clay ***chemical resistant brick*** *offer engineers a wide range of utility in environments that involve aggressive corrosive conditions and elevated temperatures."*

  Along with their chemical resistant brick their products include: Face Brick; Thin Brick; Oversized Brick; Structural Brick; Pavers; and Special Shapes.

04 70 00 - Manufactured Masonry

- **bioMASON**
  12 T. W. Alexander Drive
  Bldg. 200, Suite B
  Box 152
  Research Park, NC 27709
  Phone: 919-977-3841

  http://biomason.com/

  "bioMASON employs natural systems as industrial manufacturing, eliminating waste while providing virtually limitless variation" . . . [to **GROW BRICKS** which are] *Made in ambient temperatures, no CO2 emissions, minimal dependency on natural fossil fuels, and can be made on site, [and offer] No waste in manufacturing."*

C2C Products Innovation Institute 1st Place winner!
http://inhabitat.com/meet-the-winners-of-the-first-cradle-to-cradle-product-innovation-challenge/

- ### Vintage Thin-Sliced Salvaged Brick

Vintage Brick Salvage, LLC
1303 Harrison Avenue
Rockford, IL 61104
Phone: 800.846.3652

http://bricksalvage.com/

Vintage Brick Salvage LLC., cuts and repurposes **thin-sliced antique brick tiles** for various wall or flooring applications, but also offers salvaged / reclaimed common building brick and street pavers.

04 72 00 - Cast Stone

- ### The Maintenance, Repair, and Replacement of Cast Stone

National Park Service
US Department of the Interior
Technical Preservation Services
Preservation Brief 42

https://www.nps.gov/tps/how-to-preserve/briefs.htm

- ### Cast Stone Institute

813 Chestnut Street
PO Box 68
Lebanon, PA 17042
Phone: 717-272-3744
Fax: 717-272-5147

http://www.caststone.org/

"Cast Stone is a masonry product that provides ornamental or functional features to buildings and other structures. Cast Stone is made from fine and coarse aggregates, Portland cement, mineral oxide color pigments, chemical admixtures and water. Not surprising then, Cast Stone products are available in virtually any color, and will give the appearance of a variety of natural building stones including but not limited to limestone, granite, slate, travertine or marble. Applications for Cast Stone range from the simplest windowsills to the most complicated architectural elements."

"The difference between Cast Stone and Adhered Manufactured Stone Masonry Veneer" (Cultured Stone)
http://www.caststone.org/pdf/CastStone_vs_AdheredVeneer_9_2008.pdf

For a selection of **04 72 00** related LEED products, see:
http://www.arcat.com/divs/sec/sec047200.shtml

Masonry: *Some additional information*

- **Papercrete** (Barry J. Fuller, Tempe, AZ)

  http://www.livinginpaper.com/
  Mr. Fuller offers courses, How-To videos, and other information related to his success in recycling newspaper, junk mail and magazines to create an innovative building block.

  Paper Palace One - is one of his projects:
  https://www.youtube.com/watch?v=eM34F4ZSmXQ 2:55

  *(Papercrete would be considered an Alternative Material. See **Division 13** - for a list of other **Alternative Materials** for **Alternative Construction** based projects.)*

- **Stone or Rubble Foundations** (Existing construction)
  http://buildingscience.com/documents/insights/bsi-041-rubble-foundations

  *(See **Division 13 - Rubble Home** Alternative Construction topic)*

- *See CSI Section **09 63 40 - Stone Flooring,** for related products and information .*

End of Division 04

Space for notes:

DIVISION 05 - Metals

Before 'green was green', various metals have historically been and continue to be recycled, providing a *continuous loop* to repurpose a products' previous use into a new product.

For more information on the following, see:

- **American Iron & Steel Institute**
  25 Massachusetts Avenue, NW Suite 800
  Washington, DC 20001
  Phone: 202.452.7100
  http://www.steel.org/Sustainability/Steel%20Recycling.aspx

- **The Aluminum Association**
  1525 Wilson Boulevard, Suite 600
  Arlington, VA 22209
  Phone: 703.358.2960
  http://www.aluminum.org/sustainability/environmental-product-declarations

- **Copper Development Association, Inc.**
  260 Madison Avenue, New York, NY 10016
  Phone: 212.251.7200
  Fax: 212.251.7234
  Email: questions@copperalliance.us
  http://www.copper.org/environment/green/

It is important to remember, that when two dissimilar metals come in contact with each other, there can be **galvanic action**, causing corrosion to occur. This could simply be the obvious when two different types of metal are attached together. But, something that isn't considered is when certain fasteners (nails, screws and other hardware), are used with ACQ Pressure Treated Lumber. A similar problem happens when wrapping aluminum coil stock around this same lumber. This corrosion is a result of using high amounts of copper as part of the wood preservative. *(See **https://www.buildloghomes.org/the-corrosive-problem-acq-treated-lumber/** and **Division 06** for related addition information.)*

05 00 00 - Metals

*- See CSI Section **03 20 00 – Concrete Reinforcing**, for various reinforcing products.*

*- See CSI Section **04 05 23 - Masonry Accessories**, for hot dipped galvanized products, including steel lintels;*

wall ties; anchor bolts; metal lath, and other metal reinforcing products for masonry.

- Note: Hot-Dip Galvanized metal or Stainless Steel should be evaluated for use, per application. See:

- **American Galvanizers Association**
  6881 South Holly Circle, Suite 108
  Centennial, Colorado 80112
  Phone: 720-554-0900
  Fax: 720-554-0909
  email: aga@galvanizeit.org
  www.galvanizeit.org

- **Specialty Steel Industry of North America** (SSINA)
  The Stainless Steel Information Center
  3050 K. Street, N.W.
  Washington, DC 20007
  Phone: 202.342.8630
  http://www.ssina.com/index2.html

  LEED™ Fact Sheet for Stainless Steel:
  http://www.ssina.com/leed.htm

05 12 00 - Structural Steel Framing

- **American Institute of Steel Construction**
  One East Wacker Drive, Suite 700
  Chicago, IL 60601-1802
  Phone: 312.670.2400
  Fax: 312.670.5403

  https://www.aisc.org/

  AISC provides membership at different levels relating to steel design, detailing, manufacturing and steel erection, for professionals, educators and students

  Chicago City of Steel: *A self-guided walking tour celebrating Chicago's Structural Steel Heritage*
  http://media.aisc.org/Files/Chicago_Walking_Tour_v2011.pdf

- **Banker Steel**
  Corporate Offices and Production Facility 1
  (Main Plant)
  1619 Wythe Road
  Lynchburg, VA 24501
  Phone: 434.847.4575

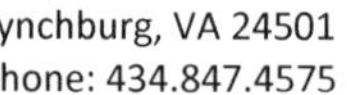

  http://www.bankersteel.com/about-banker-steel/ - with 5:14 video

  Banker Steel . . .*"employ(s) a LEED Accredited Professional Engineer on staff and we use American Steel from 91% recycled content."*

With (2) production facilities in VA and (1) in Florida, distribution of Banker Steel includes all states east of the Mississippi River.

- ## Jersey Shore Steel

P.O. Box 5055
70 Maryland Avenue
Jersey Shore, PA 17740 USA
Phone: 570.753-3000
Toll-Free: 800-833-0277
Fax: 570.753.3782
http://www.jssteel.com/green-steel with 4:49 video

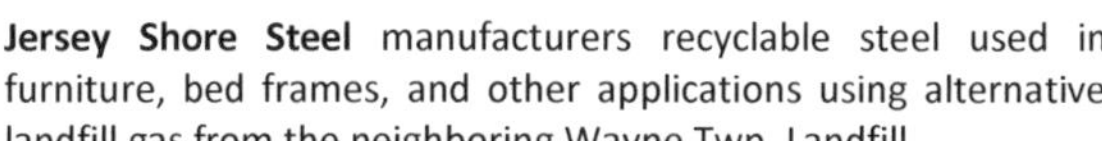

Jersey Shore Steel manufacturers recyclable steel used in furniture, bed frames, and other applications using alternative landfill gas from the neighboring Wayne Twp. Landfill

05 15 00 - Wire Rope Assemblies

- ## Associated Wire Rope Fabricators

28175 Haggerty Road
Novi, MI 48377
Phone: 248-994-7753
Fax: 800-444-2973

https://awrf.org/

"While AWRF stands for Associated Wire Rope Fabricaators, membership covers the spectrum of the lifting, rigging and load securement industry. Membership is extend to manufacturers and distributors alike."

- ## Bethlehem Wire Rope

Wirerope Works, Inc.
100 Maynard Street
Williamsport, PA 17701
Phone: 800-541-7673
Fax: 570-327-4274

http://www.wireropeworks.com/

*"**Wirerope Works, Inc.** manufactures Bethlehem Wire Rope®, the trade name under which we produce, sell and service our wire rope and strand products. The name "Bethlehem Wire Rope" represents the most complete facility and experienced personnel in North America. Our 46-acre manufacturing complex in Williamsport, Pennsylvania, with over 620,000 square feet under roof, is the single largest wire rope manufacturing facility in North America. Wirerope Works, Inc. manufactures its own wire, wire rope, structural strand, and all fabricated products such as pendants and other assemblies on the same premises."*

With other high-wire successes elsewhere, **Nic Wallenda**, crossed Niagara Falls, in June, 2012 - using a 2" dia. wire/cable from **Wirerope Works, Inc.** See the below for more info:
http://wnep.com/2013/06/25/wallenda-skywire-rope-made-in-williamsport/

For additional **05 15 00** related LEED products, see:
http://www.arcat.com/divs/sec/sec051500.shtml

05 20 00 - Metal Joists

- ### Hambro Structural Systems

  Canam - Hambro Buildings
  450 East Hillsboro Blvd.
  Dearfield Beach, FL 33441
  Phone: 954.571.3030

  https://www.canam-construction.com/en/construction-products/composite-floor-system/

  The **Hambro D500 Composite Floor System** promotes maximum duct openings, long spans, fast & easy installation, fire protection, the ability to reuse plywood forming and uses less concrete & reinforcement.

  https://www.canam-construction.com/wp-content/uploads/2014/12/canam-hambro-brochure.pdf

  Discover "The Hambro Way" :
  https://www.youtube.com/watch?v=WPY8O WHeXW0 6:35

- ### Vulcraft Steel Joists (A Nucor Company)

  Contact: Bob Wesley
  National Accounts Sales Manager
  Nucor Vulcraft Group
  Phone: 678.965.6667

  http://www.vulcraft.com/ *(with BIM related info)*

  Vulcraft has manufacturing plants in South Carolina, Nebraska, Alabama, Texas, Indiana, Utah and New York, and it has sales offices across the USA providing engineering, sales and support for **open-web steel joists, Eco-Span Composite Floor System**, along with **Steel Roof & Floor Deck** - using **BIM add-in down loadable technology.**

05 30 00 - Metal Decking

- ### Steel Deck Institute

  2661 Clearview Rd., #3
  Allison Park, PA 15101
  Phone: 847-458-4647

  http://www.sdi.org/

  "The Steel Deck Institute was formed to bring uniformity to the design, manufacture, quality control and construction practices applicable to cold-formed steel decking".

  SDI offers membership and a list of members that produce Steel Deck and a list of associate members that produce related products.

- ### Cordeck

  Headquarters

12620 Wilmot Rd.
Kenosha, WI 53142
Phone: 877.857.6400

http://www.cordeck.com/

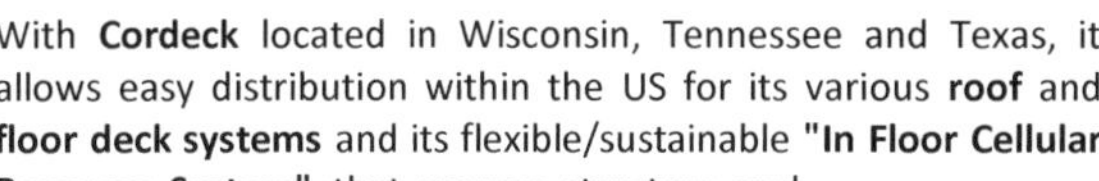

With **Cordeck** located in Wisconsin, Tennessee and Texas, it allows easy distribution within the US for its various **roof** and **floor deck systems** and its flexible/sustainable **"In Floor Cellular Raceway System"** that merges structure and wire management.

Coredeck - In Floor Raceway System
https://www.youtube.com/watch?v=N7_O4F90zT4#t=107 2:55

- ## Roof Deck, Inc.

  80 Twin Rivers Drive West
  Hightstown, New Jersey 08520
  609.448.6666
  http://roofdeckinc.com/

05 40 00 - Cold-Formed Metal Framing

This section is for **structural metal framing** (i.e. Cold-Formed steel studs and joists).

*- See **Division 09** . . . for <u>non-structural</u> metal framing.*

*- See **Division 06** . . . Metal Fasteners for Wood, Plastics and Composite lumber.*

- ## ClarkDietrich Building Systems

  Corporate Office
  9100 Centre Pointe Drive, Suite 210
  West Chester, OH 45069
  Phone: 513.870.1100
  Fax: 513.870.1300

  http://www.clarkdietrich.com/

  ClarkDietrich engineers and manufacturers **cold-formed structural steel framing** products and systems, including but not limited to **joists, studs, BIM Services,** and **LEED Services** for projects of all sizes.

- ## Steeler Construction Company

  Corporate Office
  10023 MLK Jr. Way South
  Seattle, WA 98178
  Phone: 800.275.2279
  Local: 206.725.2500
  Fax: 206.725.1300

  http://www.steeler.com/

  Steeler Construction Company *"manufactures and distributes high quality steel framing products for just about any project need"* - that can provide LEED credits as with the ELITE Steel Framing System™.

Other Steeler locations include: Tacoma, WA; Spokane, WA; Portland, OR; Tigard, OR; Sacramento, CA; San Francisco, CA; Bakersfield, CA; San Diego, CA; Phoenix, AZ; Tucson, AZ; and Delta, BC.

The **International Residential Code** (IRC) Sections that could apply, include:

- **R505** - *Steel Floor Framing*
- **R603** - *Steel Wall Framing*
- **R804** - *Steel Roof Framing*

05 50 00 - Metal Fabrications

- **Clark Steel Fabricators, Inc.**
  12610 Vigilante Road
  P.O. Box 1370
  Lakeside, CA 92404
  Phone: 619.390.1520

http://www.clarksteelfab.com/

"Since 1977, Clark Steel Fabricators, Inc. has been fabricating sizable and prestigious displays of grand architecture, and continues today reshaping San Diego's skyline with precision and quality."

- **Eisenhart Steel Company**
  17 Barnhart Drive
  Hanover, PA 17331
  Fax: 717.633.5296
  Phone: 717.633.9877
  http://www.eisenhartsteelco.com/

Eisenhart offers steel fabrication and steel erection. Contact for complete services.

- **Great Lakes Welding & Fabrication, LLC.**
  Ray Township, Michigan 48096
  Fax: 586.749.7229
  Phone: 586.201.6018
  Email: info@greatlakesweldingandfab.com

http://www.greatlakesweldingandfab.com/

Great Lakes Welding & Fabrication includes Structural Steel and custom fabrication work using Wrought Iron, Steel, Aluminum, Stainless steel, Bronze, and Brass

05 70 00 - Decorative Metal

- **Fast Fab Steel Co.**
  739 Morse Ave.
  Schaumburg, IL 60193-4533
  Phone: 847.301.7425

http://fastfabsteel.com/ornamental-iron/2786407

Along with **Ornamental Iron** work, **Fast Fab** provides engineering, fabrication and delivery of **Custom Stairs & Railings, Tube & Pipe Railings, Steel Pan Stairs, Spiral Stairs, Steel Door Jambs, Ladders** and **Cat Walks**.

- ## Reynolds Iron Works

157 Palmer Industrial Road
Williamsport, PA 17701
Phone: 570.323.4663
Fax: 570.322.4470
Inquiries: riwi@reynoldsironworks.com
http://www.reynoldsironworks.com/about/

Reynolds Iron Works provides . . . *"all types of structural steel fabrication such as beams, columns, trusses, plate fabrication, tanks and miscellaneous steel fabrication such as all types of stairs, railings, ladders and more, all available in steel, aluminum, and stainless steel."*

- ## Precision Metal Fabricators

236 39th Street
Brooklyn, NY 112200
Phone: 718.832.9805
Fax: 718.832.9405

http://www.pmfmetal.com/

*"**Precision Metal Fabricators** has been creating unique and creative metal designs for almost thirty years. We are a family owned and operated business that was established in 1988. We originally started out in the roofing materials industry but as time went on, our knowledge increased and our trade advanced to include high end commercial and residential architectural work."*

End of Division 05

Space for notes:

DIVISION 06 - Wood, Plastics and Composites

Wood. It typically is the only commonly used structural material In North America that is naturally renewable. For instance, for all the trees that grew back in and around Williamsport, PA *(The historical Lumber Capital of the World),* after being cut and logged in the late 1880's, the forests are healthy today after regenerating from seed.

In North Amercia, it is and should continue to be the goal for tree growth to exceed cutting, or otherwise known as harvesting. Unfortunately, some countries are not as blessed with the soft and hardwoods that North America has, and in some cases there was poor management practices and clear cutting has taken place.

As you know, **wood** is one of those materials that is appreciated by many. It is easy to work with and is a versatile product for building construction, furniture and a multitude of other opportunities. But, more importantly, "*Wood is Green!*".

See related brochure with the same name at: **http://keystonewoodpa.org/wp-content/uploads/2015/04/WoodIsGreenBrochure.pdf** It communicates *"10 reasons to use Pennsylvania Hardwoods For Your Next Project"*, and offers a counter perspective from what is being taught in some K-12 schools. To summarize the brochure; *Wood is . . .*

1. Renewable; 2. Environmentally-Friendly; 3. Zero Waste; 4. Carbon Neutral (and being proven to be Carbon Negative); 5. Caron Storage; 6. Green Energy; 7. Buy Local; 8. Buy American; 9. The US leads the world in Sustainable Forestry; 10. Pennsylvania leads the US in hardwood volume.

With **Wood**, **Plastics** and/or **Composites**, it should be noted that proper fasteners should be used in all applications. For example, the use of hot-dipped galvanized, stainless steel, or aluminum nails or screws on exterior wood that is exposed to the elements, would be best to prevent rust and ultimately wood rot. Through time, even decay resistant wood, such as cedar will rot if the fastener rusts. In another example, when **ACQ Pressure Treated (PT)** wood is used with the wrong fastener, corrosion and potentially severe failure can occur when the wrong fastener is used to secure it. And, if attaching metal flashing and/or aluminum break metal wrap (coil stock), such metal can corrode when attached directly to the PT wood. So for PT wood, follow manufacturers' specifications.

*(See various sections within **Division 07** and Section **10 81 00 - Pest Control Devices,** for additional flashing information)*

Plastics have come a long way, especially with the demand for single-steam recycling, yet there is plenty that ends up not being recycled. As we know, plastics do not decompose very quickly, or at all, be it in the landfills or our oceans, as revealed with the **Great Pacific Garbage Patch / Pacific Gyre**. Plastics have become important in our lives, yet if exposed to the sun, some may fade, and some may become brittle and break. As imagined, it means it is time for disposal. It is at this point we hope to recycle it, if at all possible. The problem though lies with the those products that do not end up being recycled, along with the air pollution that is generated from the manufacturing of virgin plastic, the recycling of plastics, and the burning/incinerating of plastics. The United States has various regulations, but it is unfortunate that some other countries do not - as communicated in the news.

Composite Lumber, commonly used for decking is typically a combination of wood fiber, plastics, plus a binder. As with anything, be careful and do your research, as there are composite lumber products that have had issues and lawsuits for flaking, fading, mold growth, cracking and premature failures. Such unfortunate results really mean real environmental costs to produce the product then remove and disposal of such within a short period of time. This is NOT sustainable.

Wood however, is not only naturally beautiful, it is a renewable building material and typically understood to be the most sustainable of all materials as it provides carbon negative qualities as it absorbs CO_2.

For more information see:

- **Great Southern Wood Preserving Inc.**
  P.O. Box 610
  Abbeville, AL 36310
  Phone: 334.585.2291
  Fax: 334.585.4353
  http://www.greatsouthernwood.com/products/fastenerinfo

- **Simpson Strong-Tie**

Home Office:
P.O. Box 10789
Pleasanton, CA 94158
Phone: 800.925.5099
Fax: 925.847.1597

http://www.strongtie.com/productuse/ptwoodfaqs.html#
The above discusses CCA required hardware.

- ## American Wood Council

222 Catoctin Circle SE, Suite 201
Leesburg, VA 20175
Phone: 202.463.2766
Fax: 202.463.2791

http://www.americanwoodcouncil.org/

- ## American Wood Protection Association

Formerly - American Wood Preservers' Association
Mailing Address:
P.O. Box 361784
Birmingham, AL 35236-1784

Shipping Address:
100 Chase Park South, Suite 116
Birmingham, AL 35244-1851
Phone: 205.733.4077
Fax: 205.733.4075
http://www.awpa.com/

- ## Architectural Woodwork Institute

46179 Westlake Drive, Suite 120
Potomac Falls, VA 20165
Phone: 571.323.3636
Fax: 571.323.3630
General Inquiries: info@awinet.org
Technical Inquiries: help@awinet.org

http://www.awinet.org/

- ## Forest Stewardship Council (FSC)

212 Third Avenue North, Suite 445
Minneapolis, MN 55401
Phone: 612.353.4511

https://us.fsc.org/

*See CSI Section **06 11 01 - FSC Certified Wood Framing Lumber**, for more information.*

- ## Hardwoods Development Council

Pennsylvania Department of Agriculture
2301 North Cameron Street
Harrisburg, PA 17110
Phone: 717.772-3715

http://www.agriculture.pa.gov/Pages/search.aspx?searchBox=hardwoods%20council#.VsPxgogo6P8

The Pennsylvania Department of Agriculture / Hardwoods Development Council welcomes the opportunity to schedule a presentation (that offers AIA credits) and to gain knowledge about PA's forests and its lumber. With this, request to have the **'WoodMobile'** exhibit trailer brought to your location for hands-on learning, for all ages.

- **National Hardwood Lumber Association**
  PO Box 34518
  Memphis, TN 38184
  Phone: 901.377.1818
  Member Hotline: 800.933.0318

  http://www.nhla.com/

- **Northern Tier Hardwoods Assoc. (NTHA)**
  P.O. Box 7
  Tunkannook, PA 18657
  Phone: 570-265-7753

  http://www.nthardwoods.org/

- **The Engineered Wood Association**
  Formerly - American Plywood Association (APA)
  7011 S. 19th Street
  Tacoma, WA 98466-5333
  Main: 253.565.6600
  Product support: 253.620.7400
  Fax: 253.565.7265

  http://www.apawood.org/

- **Pennsylvania Sustainable Forest Initiative**
  211 Barrington Lane
  Bellefonte, PA 16823
  Phone: 814.355.1010
  Fax: 814.355.1022

  http://www.sfiofpa.org/

- **Plastic Lumber Trade Association** (PLTA)
  P.O. Box 211
  Worthington, MN 56187
  Phone: 507.372.5558
  Fax: 507.372.5726

  http://www.plasticlumber.org/

- **Sustainable Forestry Initiative** (SFI)
  2121 K Street, NW, Suite 750
  Washington, DC 20037
  Phone: 202.596.3450
  Fax: 202.596.3451

  http://www.sfiprogram.org/

  *See CSI Section **06 11 01 - FSC Certified Wood Framing Lumber**, for more information.*

- **Viance - Treated Wood Solutions**

8001 IBM Drive, Bldg. 403
Charlotte, NC 28262
Phone: 800-421-8661
Fax: 714-527-8232

http://treatedwood.com/about-us/

Viance Treated Wood Solutions is the manufacturer of wood for exterior applications and those for building code fire-retardant purposes, known as: "*D-Blaze*® *Fire Retardant Treated Wood (FRTW). [D-Blaze] is now* ***GREENGUARD GOLD*** *certified for low chemical emissions.*"

- **Wood Products Manufacturers Association**

P.O. Box 761
Westminster, MA 01473-0761
Phone: 978.874.5445
Fax: 978.874.9946

http://www.wpma.org/pages/home/

06 00 00 - Construction Adhesives

Construction adhesives have come a long way over the years by reducing or eliminating **Volatile Organic Compounds** (VOCs) that historically have caused poor **Indoor Air Quality** (IAQ), and contributed to what has been known as **Sick Building Syndrome**. The below offer either low-VOC and/or No-VOC products.

- **DriTac**

60 Webro Road
Clifton, NJ 07012
Phone: 973-614-9000
Fax: 973-614-9099

http://www.dritac.com/family-green-adhesives/

"***DriTac Flooring Products*** *is pleased to feature our family of "Green" Environmentally Friendly Adhesives and Installation Products for the Flooring Industry. DriTac recognizes the importance of the "Green" movement.*

We are actively working with architects, builders and developers to address this market need. You can rely on DriTac to provide you with high quality adhesives for your flooring needs while contributing to your "green" building efforts!

DriTac produces a full line of zero VOC, zero solvent Certified Green flooring installation solutions."

- **Titebond**

Franklin International
2020 Bruck Street
Columbus, OH 43207
Phone: 800-877-4583
Fax: 614-445-1295

http://www.titebond.com/index.aspx

"Every day, woodworkers, professional contractors and do-it-yourselfers reach for the brand they trust most. ***Titebond.*** *Our brand promise of unsurpassed product quality and technical expertise makes Titebond a preferred global choice."*

Note: There may be other adhesives that are Low or No-VOC, while also being made in the USA, but could not be validated in time for publishing.

06 01 00 - Maintenance of Wood, Plastics and Composites

As mentioned elsewhere in the book, all products and materials need some level of maintenance. And, it is no different with Wood, Plastics and Composites.

Plastic and Composite products are commonly a unique to a particular company. If this product is no longer manufactured for whatever reason, the section that is damages is then hard to be replaced or repaired. To be able to repair only the damaged portion is the sustainable option.

What is nice about wood is that it a section can be replaced, shaped and/or restored quite easily as presented by the information below:

- **Abatron, Inc.**
  5501 95th Ave.
  Kenosha, WI 53144
  Toll Free: 800-.445.1754
  Phone: 262.653.2000
  Fax: 262.653.2019

http://www.arcat.com/arcatcos/cos30/arc30046.html?qstr=06 0100&pids=115337 136783 with videos and LEED info

*"**Abatron, Inc.** formulates, manufactures, and sells an extensive line of products for building restoration, DIY repairs, potting and encapsulating, OEM, casting and moldmaking, composites, and many other applications. Popular items include epoxy wood repair products, concrete repair products, structural adhesives, and protective coatings. Founded in 1959, the company excels at producing high-performance products supported by a trained technical staff and easily accessible customer support."*

- **Preserving Historic Wooden Porches**
  US Department of the Interior
  Technical Preservation Services
  Preservation Brief 45

https://www.nps.gov/tps/how-to-

preserve/briefs.htm

- **Protective Coating Company**
  221 S 3rd Street
  Allentown, PA 18102
  Toll Free: 800.220.2103
  Fax: 610.432.5403

  http://www.pcepoxy.com/

  Protective Coating Company is *"ccelebrating 60 years of Fixing Your Things! Raise your glass to Made in USA and the Fix Your Things movement!* To do so, **Protective Coating Company** is known for its ***PC-Petrifier®*** and ***PC-Woody®*** products for restoring various projects or products made from wood.

Additional information can be found in the **International Residential Code** (IRC), ***Section R317 - (Protection of Wood and Wood Based Products Against Decay)***, and *Section* ***R318 - (Protection Against Subterranean Termites)***.

Use this website for *PremiumACCESS or* free *PublicACCESS* viewing of ***I-Codes*** *and* ***ICC Standards***: **https://codes.iccsafe.org/**

06 09 00 - Wood and Plastic Fastenings

- **Maze Nails**
  100 Church Street
  Peru, IL 61354
  Toll Free: 800.435.5949
  Local: 815-223-8290
  Fax: 815.223.7585

  http://www.mazenails.com/

  America's largest manufacturer of specialty nails, including roofing, siding, trim, decking, gutters, spikes, fencing, pole barn and other applications.

  Note: It is understood that **Maze Nails** is the manufacturer of nails used in the ***"All-American Home"*** described at the front of the book, built recently by **Anders Lewendal** and crew - in Bozeman, MT.

- **Simpson Strong-Tie**
  Mailing Address:
  PO Box 10789
  Pleasanton, CA 94588

  Street Address:
  5956 W. Las Positas Blvd.
  Pleasanton, CA 94588
  Local Phone: 925.560.9000

Toll-Free: 800.925.5099
Fax: 925.847.1597

http://www.strongtie.com/#

Simpson Strong-Tie provides a full line of structural Connectors, Fastening systems, Anchoring systems, Repair, Protection & Strengthening systems, products and services for sustainable design and construction opportunities - including, but not limited to Advanced Framing techniques, Seismic, Wind and other important structural advances.

Because tall, platform framed construction has gained in popularity, **The NEESwood Capstone Test Building** was funded by the **National Science Foundation**, and led by Colorado State University and Simpson Strong-Tie, was performed at the **Hyogo Earthquake Engineering Research Center**, Miki City, Japan. **It** demonstrated the **World's Largest Earthquake Test** showing a 7 story, 23,000sf, 800,000lb., wood framed building that was mounted to a test table that moves simultaneously, in 3 dimensions. The results validate and will help future performance-based structural design strategies that should be incorporated, to build such structures within various seismic zones.

World's Largest Earthquake Test:-
https://www.youtube.com/watch?v=9X-js9gXSME 6:27

- **Tree Island - (Halsteel and True Spec Nails)**
  4190 East Santa Ana
  Ontario, CA, 91761
  Phone: 800. 255.6974

  http://www.treeisland.com/brands/halsteel

http://www.treeisland.com/brands/truespec

Tree Island offers quality American made wire and wire products since 1964, including nails for various applications, such as the ZincGard® hot dipped galvanized nails for ACQ Treated Lumber, or other common and specialty applications.

- **Tremont Nail Company**
  457 School Street
  Mansfield, MA 02048
  Phone: 508-339-4500
  Fax: 508-339-0104

  http://www.tremontnail.com/

  Tremont Nail Co. offers over 190 years of excellence, manufacturing *"Steel Cut Nails for Authentic Restoration Projects & Remodeling"*.

06 10 00 - Rough Carpentry

Rough Carpentry in Divison 06 relates to the wood 'bones' of the building, also known as **Light-Frame Construction**. It can include the floor, wall and roof framing, where specified, or exists in an existing buidling.

Historically, these wood framing members (floor joists, wall studs, ceiling joists, roof rafters, beams, etc.) are solid sawn wood members and when originally sized, they were rough sawn to the common sizes of 2"x4", 2"x6", 2"x8", 2"x10" and 2"x12" and typically were a variety of hard and soft wood species that were grown, cut and milled locally. (See the following link for ***"History of Yard Lumber Size Standards"***, by the US Dept. of Agriculture.
http://www.fpl.fs.fed.us/documnts/misc/miscpub_6409.pdf

Today, these common sizes are similar, but after being surfaced on four sides (S4S), they are known by the same names, but are typically labeled as 2x4, 2x6, etc. This would represent the nominal vs. the actual size. Commonly, on an architectural drawing, if the material is labeled as 2x6, it represents the nominal size. If labeled 2"x6", it represents the material should be 2"x6" in size. Many times, this can be achieved if the 2"x6" is rough sawn. (i.e. 2"x6" RS)

The related dimension(al) lumber sizes (nominal and actual) are illustrated at:

http://www.ezwoodshop.com/lumber-dimensions.html

When deciding on a structural floor, wall and/or roof system, it is wise to compare all the material and product options when considering the potential of fire, hurricanes, tornadoes, earth quakes, flooding, and other natural disasters.

If it was decided to use dimension(al) framing lumber, it is important to remember that when drawing and building with such, always draw the materials actual size (drawn to scale), and label accurately. (i.e. 2x6 studs at 24"o.c.) This would mean that 2x6 (1.5"x5.5") wood studs are to be spaced 24" on-center, to create the wall framing.)

Advanced Framing, also known as **Optimum Value Engineering** was developed in the 1970's by the **National Association of Home Builders (NAHB) Research Center.** *(Yes, you read that right - 1970's!)* This technique typically considers ways to reduce the number of redundant wood framing members, while keeping the design structurally sound, less labor, and other potential benefits.

Advanced Framing: Meet Structural Code & Energy Requirements -
https://www.youtube.com/watch?v=ehGmQ3KWxAE 4:25

The same video, plus free Advanced Framing CAD Details and other info., is found at APA. (previously known as American Plywood Assoc. and is currently known as The Engineered Wood Assoc).

See: **http://www.apawood.org/walls** (Then, scroll to bottom)

Or, the **APA Advanced Framing Construction Guide** can be accessed without registration at:

http://www.apawood.org/data/sharedfiles/documents/m400.pdf

Per Green Building Advisor, see the Pro's and Cons to Advanced Framing:
http://www.greenbuildingadvisor.com/blogs/dept/musings/pros-and-cons-advanced-framing
(if problems, search ' *Pro's and Con's to Advanced Framing - Green Building Advisor* ')

In theory, it Advanced Framing is great. It minimizes wood, labor and related costs. But, to achieve it can be a challenge, especially when standard door and window sizes do not match the framing module. If a little door and window research is done, and layout a wall showing 24" stud spacing is created, this is evident. For instance, to install a window or door in a 22.5" stud bay, this would not meet the egress requirements for a door or window. And, if a window or door was installed in the space between two stud bays (22.5" + 1.5" + 22.5" = 46.5"). Now, in this case, this would be custom door width. To reduce this rough opening width to fit a standard 36" door would be the lesser cost option, but requires additional stud(s), going against this theory. Right?!

In another example, eliminating one of the top plates is also a concern when considering drywall sizes. As you know, a standard ceiling traditionally has a 97-1/8" height, (one bottom plate, the pre-cut studs, and the two top plates), before the ceiling drywall and any finish flooring is installed. Do the math and see how labor is added to trim the drywall, while also creating more waste. A problem. ☹

Rough Carpentry / Wood Framing: *Some additional information -*

- How to Design A Wall - (Describes 2 popular approaches) **http://www.greenbuildingadvisor.com/articles/dept/musings/how-design-wall** (if problems, search, ' *How to Design a Wall* '.)

- Larry & Joe Haun: Taunton Press Wood Framing Series - **https://www.youtube.com/watch?v=dqM7ekLfSmA&list=PL4B8173887CB4CD24** (multiple videos for the series)

- **Lumber grades and types**
  http://www.thomasnet.com/articles/plant-facility-equipment/lumber-types

  This website gives definitions to the industry driven Lumber Grades and Lumber Types. It also provides links to specific hardwood and softwood suppliers.

- **Great Northern Lumber**
  2200 W. 127th St.
  Blue Island, IL 60406
  Phone: 888-586-5523
  Fax: 708.388.0887

  http://greatnorthernlumber-px.rtrk.com/gnl/

  Per ThomasNet.com - Great Northern Lumber is a *"manufacturer and wholesale distributor of industrial lumber, plywood and panels. Manufacturer of wooden concrete forms. Various capabilities include vendor inventory management, yard distribution, shrink wrapping, heat treating, consulting, custom cutting and remanufacturing. Aerospace, building and heavy-construction, commercial and industrial applications served. FSC certified. Same day delivery."*

- **Weyerhaeuser**
  33663 Weyerhaeuser Way
  South Federal Way, WA 98003
  Local: 253.924.2345
  Toll free: 800.525.5440

  http://www.woodbywy.com/lumber/weyerhaeuser-green-studs/

  Manufactures and distributes Rough Carpentry products, including: Spruce, Pine, Hem-fir, Southern Yellow Pine, Douglas Fir 'Green Studs', and various Oriented Strand Lumber.

The International Residential Code (IRC) sections that apply to Rough Carpentry / Wood Framing include:
- **R502** - *Wood Floor Framing*
- **R602** - *Wood Wall Framing*
- **R802** - *Wood Roof Framing*

06 11 01 - FSC Certified Wood Framing Lumber

What is the difference between the **Forest Stewardship Certification (FSC)** and the **Sustainable Forest Initiative (SFI)**?

It is interesting that CSI provides a division number for FSC, but one for SFI is unfound. It is also interesting that FSC is accepted by United States Green Building Council (USGBC) and SFI is not. But, why? Refer to the below websites to help clarify the difference and debate.

- **FSC vs. SFI what is the difference?**
  http://www.snwwood.com/Blog/Wood-Q-A/FSC-vs-SFI-What-s-the-difference
  (if problems, search '*FSC vs. SFI what is the difference?*')

- **https://www.houselogic.com/remodel/remodeling-tips-advice/what-is-fsc-certified-wood/**
  (if problems,, search ' *What is FSC certified wood, House Logic* ')

- **Forest Stewardship Council** (FSC - US)
  212 Third Avenue North, Suite 445
  Minneapolis, MN 55401
  Phone: +1 612.353.4511

  https://us.fsc.org/en-us

- **FSC Certificate Database**
  <u>Use the database to:</u>

  1) Verify if a company is FSC certified.
  2) Search for FSC certified companies or products.

  https://us.fsc.org/fsc-certificate-database.311.htm

- **Sustainable Forest Initiative, Inc.** (SFI)
  2121 K. Street, NW, Ste. 750
  Washington, DC 20037
  Phone: 202.596.3450
  Fax: 202.596.3451

  http://www.sfiprogram.org/

  The above website also provides a database of SFI Certified lumber products.

FSC vs. SFI - You have to make the decision.
Forest Stewardship Council (FSC) vs. Sustainable Forest Initiative (SFI). Not only is it important to understand if the lumber products were sustainably harvested, but it is critical to understand that the small grower may not be able to pay such 3rd party certification fees. So, one of the most important considerations should include purchasing local materials. Per **USGBC**, this typically means within 500 miles maximum, so to limit the environmental / transportation costs.

So, do you know where the wood that you hope to use is from? For instance, the Ipe lumber desired for the deck, it may be FSC, but what about the miles? If building in the USA, to purchase lumber that originates from Central or South America would violate this principle.

06 11 13 - Engineered Wood Framing Products

- **The Engineered Wood Association (APA)**
  7011 S. 19th Street
  Tacoma, WA 98466-5333
  Main: 253.565.6600
  Product support: 253.620.7400
  Fax: 253.565.7265

  http://www.apawood.org/products

  The Engineered Wood Assoc., formerly known as the American Plywood Association (APA) supports manufacturing of various wood products such as: Plywood, OSB, Concrete Form Panels, Siding, Radiant Barrier Panels, Glulam, I-Joist, Rim Board, Cross-laminated Timber, and various Structural Composite Lumber.

- **Georgia-Pacific**
  Georgia-Pacific Building Products
  133 Peachtree Street NE
  Atlanta, GA 30303
  Phone: 404.652.4000

  http://www.buildgp.com/engineered-lumber

  Georgia Pacific manufactures and distributes Engineered Wood Framing Products, including: **Wood I-Beam Joists**; **GP LAM LVL** (Laminated Veneer Lumber); and **FiberStrong Rim Board**

- **Huber Engineered Woods**
  10925 David Taylor Drive, Suite 300
  Charlotte, NC 28262
  Phone: 800.933.9220

  http://www.huberwood.com/

  This website has both Zip System and Advantech videos.

  Humber Engineered Woods manufactures and distributes Engineered Wood components, including: **Zip System; Advantech**; Huber Blue for wall and floor sheathing, and the **Huber Rim Board**.

- **Weyerhaeuser**
  33663 Weyerhaeuser Way
  South Federal Way, WA 98003
  Local: 253.924.2345
  Toll free: 800.525.5440

  http://www.woodbywy.com/trus-joist/

Weyerhaeuser manufactures and distributes Engineered Wood Framing Products, including: **TJI Joists; Timberstrand LSL** (Laminated Strand Lumber**); Parallam PSL** (Parallel Strand Lumber); and **MicroLam LVL** (Laminated Veneer Lumber)

06 12 16 - Stressed Skin Panels (Structural Insulated Panels)

Contact (SIPA) for a reputable contractor:

- **Structural Insulated Panel Association**

  P.O. Box 39848
  Fort Lauderdale, FL 33339
  Phone: 253.858.7472

  http://www.sips.org/about

  Contact Structural Insulated Panel Assoc. (SIPA) for a reputable contractor

- **Agriboard Industries**

  A Division of Ryan Development Co., LLC
  1401 Enid Drive
  Vernon, TX 76384
  Toll free: 866.247.4267
  Phone: 580.305.2491

  http://www.agriboard.com/index.html

  Agriboard is a Structural Insulated Panel (SIP) that is made with formaldehyde free OSB panels and highly compressed wheat or rice straw, a waste agricultural product that eliminates burning and tilling - allowing for a negative carbon footprint.

  Agribord -
  http://www.youtube.com/watch?v=txdbL8COxZs 1:31

- **Ecovative Design**

  Mycoboard *(See CSI* ***Section 06 04 00*** *for more information)*

  Eco-HQ
  70 Cohoes Avenue, Suite 103
  Green Island, NY 12183

  Eco-East
  79 102nd St
  Troy, NY 12180

  Phone: 518.273.3753
  http://www.ecovativedesign.com/

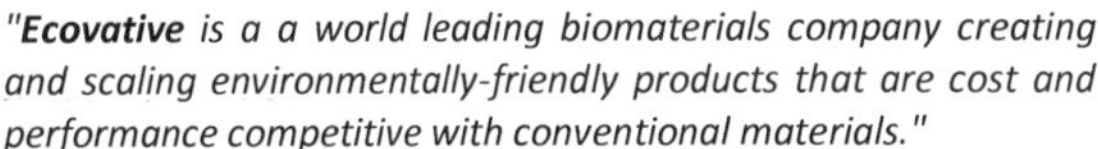

  *"**Ecovative** is a a world leading biomaterials company creating and scaling environmentally-friendly products that are cost and performance competitive with conventional materials."*

  Ecovative uses and promotes Greensulate - a Mycelium Mushroom based insulation which is said to use 8 times less energy and 10 times fewer CO2 emissions to manufacture and distribute comparable to polystyrene and polyurethane foam insulation.

  Per their website, the various Mycelium based products are also cost competitive; Home compostable; Rapidly renewable; Naturally fire resistant, VOC free, Non-Petroleum based Mouldable; and Buoyant. Other current products include Myco Board (a replacement for particleboard, plywood and fiberboard that does not use harmful resins); and Myco Foam (an earth friendly alternative that replaces foam based insulation for the packaging industry and other applications including insulation for thermal and acoustical benefits.)

http://www2.buildinggreen.com/blogs/greensulate-fungus-based-insulation-material-thats-grown-rather-manufactured

- ## Insulspan

9012 East US Hwy 223
Blissfield, MI 49228
Phone: 517.486.4844
Fax: 517.484.2056

http://www.insulspan.com/

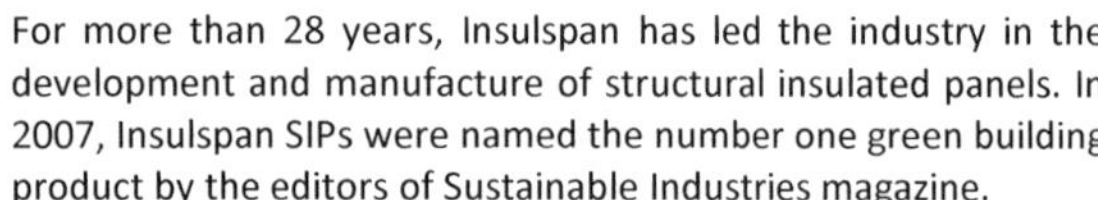

For more than 28 years, Insulspan has led the industry in the development and manufacture of structural insulated panels. In 2007, Insulspan SIPs were named the number one green building product by the editors of Sustainable Industries magazine.

Insulspan also markets the **Sustainable Affordable Kustom (SAK™) House** building packages (created by Green Built Consultants, Inc.) for living and/or working environments.

http://sakhouse.com/

- ## The Muras Company, Inc.

P.O. Box 220
3234 Route 549
Mansfield, PA 16933
Toll Free: 800.626.8787
Local: 570.549.2100
Fax: 570.549.2101

http://www.murus.com/

The **Murus Company**, Inc. offers a variety of Structural Insulated Panel (SIP) options including those made with Polyurethane (PUR), Expanded Polystyrene (EPS) and NEOPOR® (ESP) foam.

NEOPOR® is a registered product of BASF that has graphite added to the ESP, increasing the R-value, by reflecting heat.
http://www.neopor.basf.us/

06 13 13 - Log Construction

- ## The Preservation and Repair of Historic Log Buildings

National Park Service
US Department of the Interior
Technical Preservation Services
Preservation Brief 26

https://www.nps.gov/tps/how-to-preserve/briefs.htm

- ## NAHB Log and Timber Homes Council

1201 15th Street NW

Washington, DC 20005
Phone: 800-368-5242
Fax: 202-266-8400

https://www.nahb.org/en/consumers/home-buying/types-of-home-construction/types-of-home-construction-log-homes.aspx

06 13 23 - Heavy Timber Construction

Heavy Timber frame is one of the oldest types of construction practices in the US, creating the 'bones' of a building. With it, StrawBale infill and/or Structural Insulated Panels (SIPS) can provide the insulation system for which the 'skin' will be applied, for both residential and non-residential applications.

- **American Wood Council**
  222 Catoctin Circle SE, Ste. 201
  Leesburg, VA 20175
  Phone: 202.463.2766

  http://www.awc.org/codes-standards/publications/wcd5

- **Bensonwood**
  6 Blackjack Crossing
  Walpole, NH 03608
  Phone: 877.203.3562

  http://bensonwood.com/

  *"Over the years, **Bensonwood** has developed an approach to solve particular problems related to construction. Most home construction is done on site, where raw materials are delivered and workers then custom-cut each piece to build a home. This laborious process is not much different than it was one hundred years ago. We refer to our three-part solution as "**Montage Building**," derived from the photographic term meaning, "assembly."*

- **Woodhouse ®**
  3295 Route 549
  P.O. Box 219
  Mansfield, PA 16933
  Toll Free: 844.232.7339
  Phone: 570.549.6232
  Fax: 570.549.6233

  http://www.timberframe1.com/

  *"Although each custom **Woodhouse®** timber frame home is unique, there are certain elements they all share. These include our unrivaled expertise in joinery, custom home construction, our unmatched design experience, our incredible attention to detail and the completeness of our package. Plus, our unique lifetime transferable warranty."*

06 17 19 - Cross Laminated Timber (CLT)

Cross Laminated Timber (CLT) is the new kid on the block, making its way to the US after being used in Europe. And, through various tests it is shown to be a direct competition to concrete, masonry and steel structural materials.

CLT consists of several layers of kiln-dried lumber stacked in alternating directions, similar to plywood, but yet on a much larger scale.

- **SMARTLAM**
  1863 13th Street West
  Columbia Falls, MT 59912
  Phone: 406-892-2241

  http://www.smartlam.com/about/

  *"**SmartLam, LLC** is proud to be the first manufacturer of **Cross Laminated Timber** products in the United States. Our focus is to develop practical, innovative, and sustainable solutions to satisfy all our customer's project requirements."*

  "CLT is the next-generation of engineered wood products. Extensively tested and widely used in Europe for the past 20 years, CLT has limitless applications for construction, industrial matting, bridging and beyond. Anything that is built with wood, steel, concrete, or masonry can be built with CLT! Because of CLT's amazing strength, rigidity and stability, it is a cost-competitive replacement for the traditional structural materials like steel, concrete, and masonry."

Cross Laminated Timber: *Some additional information*

- **CLT Basics** from APA / The Engineered Wood Association:
  https://www.apawood.org/cross-laminated-timber

- Article - ***Cross Laminated Timber: Taking Wood Buildings to the Next Level,*** by Layne Evans
  http://www.awc.org/pdf/education/mat/ReThinkMag-MAT240A-CLT-131022.pdf

- Article - ***Cross Laminated Timber Is the Most Advanced Building Material,*** by Clay Risen
  http://www.popsci.com/article/technology/worlds-most-advanced-building-material-wood-0#page-3

- ***Introducing Cross Laminated Timber (CLT) to North America:***
  http://www.woodworks.org/design-with-wood/building-systems-clt/ with 4:11 video

  (Note: YouTube video also available by same name)

06 20 00 - Finish Carpentry

Contact the below for a reputable contractor:

- **Finishing Contractors Association**

  FCA International
  1 Parkview Plaza, Suite 610
  Oakbrook Terrace, IL 60181
  Phone: 630.537.1042
  Fax: 630.590.5272
  Toll-Free: 866.322.3477

  http://www.finishingcontractors.org/

- *See* ***Division 09 Finishes*** *- for related information*

- *See* ***Division 12 Furnishings*** *- for related casework (cabinets) and countertops.*

06 40 00 - Architectural Woodwork

Architectural Woodwork typically includes the wood products and craftsmanship used to design and create finished woodwork that includes running trim (mouldings), doors, stairs and stair components, casework (cabinetry), Historic Restoration work, along with veneers, laminates and solid surface material applications.

Contact AWI for a reputable contractor:

- **Architectural Woodwork Institute (AWI)**

  46179 Westlake Drive, Suite 120
  Potomac Falls, VA 20165
  Phone: 571.323.3636
  Fax: 571.323.3630

  http://www.awinet.org/

- **Mycoboard**

  Ecovative Design
  Eco-HQ
  70 Cohoes Avenue, Suite 103
  Green Island, NY 12183

  Eco-East
  79 102nd St
  Troy, NY 12180

  Phone: 518.273.3753
  http://www.ecovativedesign.com/

  Mycoboard, a patented Mycelium product, is a substitute for particleboard and is currently used for furniture, architectural panels, door cores, cabinetry and billboards.

- ## Pioneer Millworks

1180 Commercial Drive
Farmington, NY 14425
Phone: 585.924.9970

2609 Southeast 6th Ave.
Portland, OR 97202
Phone: 503.719.4800

http://pioneermillworks.com/

Per the above website:
- Our reclaimed wood products are fully made in the USA
- Our 13-acre NY facility is solar powered
- Our kilns are heated 100% with our own scrap wood
- We are FSC chain of custody certified in three categories
- We're Green America approved
- All of our products are LEED point eligible
- We have two LEED AP staff members on staff

- ## Purebond

Columbia Forest Products
7900 Triad Center Drive, Ste. 200
Greensboro, NC 27409
Phone: 800.637.1609

http://www.columbiaforestproducts.com/product/purebond-classic-core/#

Purebond is the original decorative Hardwood Plywood that uses Soy-Based Technology while being cost-competitive and made in North America.

- ## Terrimica

Potlatch Corporation
Plant Location: 401 N. Potlatch Road
Post Falls, ID 83854

601 West First Avenue, Ste. 1600
Spokane, WA 99201
Phone: 509-328-0930
Fax: 509-327-9409

http://www.potlatchcorp.com/files/Downloads/Particleboard/terramicaweb.pdf

Terrimica is a 100% pre-consumer recycled particleboard made with no added Formaldehyde (NAF) while being FSC Certified.

- ## Vermont Natural Coatings

P.O. Box 512
Hardwick, VT 05843
Phone: 802.472.8700

http://www.vermontnaturalcoatings.com/

*"**Vermont Natural Coatings** patented PolyWhey® technology has emerged as a new category of wood finish; delivering easy-to-use products that have established the highest performance and environmental standards. We use whey protein, a byproduct of cheese making, to displace toxic ingredients found in traditional finishes. Vermont Natural Coatings PolyWhey floor and furniture finishes and our new Exterior Penetrating Stain and Water Proofer Infused with Juniper, are the choice of professionals and do-it-yourselfers alike."*

As it was hoped to include ***strawboard*** amongst this heading, it is unfortunately learned that all the products found, such as: *EnviroWall, Invotek*, and the various *Kirei* products are all be made <u>outside</u> the USA.

It is also unfortunate that trying to contact VanBeeks Custom Wood Products, Inc., to find out if their ***Sunflower Seed Board*** is still manufactured, along with where it might be made, . . . it is found that the email address is no longer valid, and the customer support by phone is non-existent. Not enough demand?

06 50 00 - Structural Plastics

- ### American Recycled Plastics, Inc.

  773 N. Union Grove Rd
  Friendsville TN 37737
  Phone: 865.738.3439
  Fax: 865.738.3731

  http://itsrecycled.com/plastic-lumber-structural/

  The **American Recycled Plastics, Inc. structural lumber** is *"manufactured with HDPE. The fiberglass elements acts as a reinforcing agent to the HDPE. All materials are colored throughout & have UV additives to prevent deterioration of the plastic lumber from exposure to UV light. This product will not rot, split, crack or splinter for a minimum of 50 years and is resistant to termites, marine borers, salt spray, oil and fungus. Its exceptional strength makes this grade of plastic lumber the construction material of choice for outdoor structural applications."*

- ### Trimax Structural Lumber

  Conrad Forest Products
  68765 Wildwood Road
  North Bend, Oregon 97459
  Phone: 800.356.7146
  Fax: 541.756.0131

  http://www.conradfp.com/building-products-trimax.shtml

  Trimax Structural Plastic Lumber is *"made of recycled plastic (HDPE) and 30% Fiber Fill, TriMax Structural Plastic Lumber is uniquely suited for outdoor applications requiring superior strength and durability. Plus, it's an ecologically sound alternative for pressure-treated lumber."* It offers *"the benefits of plastic lumber with the structural strength similar to wood."*

06 60 00 - Plastic Fabrications

- ### Fypon

  1750 Indian Wood Circle
  Maumee, OH 43537
  Phone: 800.446.3040

http://www.fypon.com/

Fypon manufactures low-maintenance cellular PVC and high-density Polyurethane moulded millwork and railing systems for various exterior applications. Call, or see website to locate a dealer.

- ## TruExterior

Boral Materials Technologies, Inc.
45 Northeast Loop 410, Ste. 700
San Antonio, TX 78216
Phone: 800.964.0951
Fax: 210.349.8512

http://www.boralamerica.com/TruExterior

Like no other, **Boral TruExterior Trim** is made from a combination of bio-based polymer chemistry and coal-combustion (ash) products that make up more than 70% recycled or rapidly renewable materials. See their website for additional information and a full range of products.

- ## Versatex Building Products, LLC

400 Steel Street
Aliquippa, PA 15001
Phone: 724.857.1111
Fax: 724.241.3599

http://versatex.com/

Versatex boards and mouldings that are designed for exterior use are made from a minimum of 10% recycled PVC. Review the Sustainability statements for further information about the company practices and products.

Is **Polyvinyl Chloride (PVC)** a green product? This long winded debate has continued for years. After reviewing the following websites, and possibly other research / documentation, ultimately you and your client will have to make the decision to use or not.

http://www.motherearthnews.com/green-homes/the-vinyl-debate.aspx

If you do dig further, you will find plenty of content that communicates the pros and cons of the chemical and leaching concerns of PVC, yet PVC products offer a certain amount of durability. **http://www.iom3.org/pvc-sustainability**

06 64 00 - Fiberglass Reinforced Panels

As typically required by the Health Department, Fiberglass Reinforced Panels (FRP) are commonly found in commercial kitchens and sometimes on public restroom walls. These panels provide a scrubable / cleanable surface. From an aesthetic point of view, they can be pretty stark looking. But they don't have to. Take a look.

- **Marlite**
  Corporate Office
  1 Marlite Drive
  Dover, OH 44622
  Toll-Free: 800.377.1221
  Phone: 330.343.6621
  Fax: 330.343.7296

http://www.marlite.com/designer-wall-systems-frp-wall-panels.aspx

"Once considered bland and boring, ***Marlite*** *now offers the industry's most innovative FRP products. Marlite FRP (fiberglass reinforced plastic) Wall Panels provide ultimate durability, satisfying the most stringent demands. Marlite FRP is tough, water-resistant, economical to install and easy to maintain."*

06 80 00 - Composite Fabrications

- **Composite Panel Association**
  19465 Deerfield Avenue, Suite 306
  Leesburg, VA 20176
  Phone: 703.724.1128
  Fax: 703.724.1588

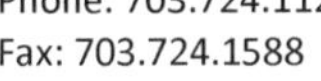

http://www.compositepanel.org/

"The ***Composite Panel Association (CPA)*** *represents the North American composite panel and decorative surfacing industries and brings together the entire value chain affiliated with composite panels."*

- **MicroLite!**
  KMDI Inc.
  400 Funston Rd.
  Kansas City, KS 66115
  Phone: 913.281.4200
  Toll-Free: 800.474.5004
  Fax: 913.281.0208

http://www.kmdi.net/

"KMDI has fashioned and installed architectural elements for many of the major designers, architects, and corporations in the United States. ***MicroLite!,*** *a proprietary material invented and produced by KMDI, is a lightweight, seamless and fully encapsulated composite material that can achieve a Class A fire rating. MicroLite! is as much a process as it is a material and as such it is especially suited to create all manner of soffits, fascia's, displays, curves, rings and clouds."*

With Wood-Plastic Composite (WPC) products designed for exterior use, it is unfortunate that such attempts to produce a product meant various class-action lawsuits on behalf of mold and fading problems. See additional comments at beginning of Division 06.

End of Division 06

Space for notes:

DIVISION 07 - Thermal & Moisture Protection

Thermal & Moisture Protection is one of the most critical of topics. This section typically addresses the exterior **building envelope**. This envelope should wrap the building with as a continuous wrap, with all penetrations carefully sealed. On a building, this wrap would include the thermal, moisture and air barriers expected for the roof, exterior walls and lowest floor of a building - enclosing the areas/spaces to be 'conditioned', those that are heated and/or cooled within the buildings' **thermal envelope**.

It should be noted that the proper **exterior flashing** is critical to prevent rain water infiltration. Poor flashing leads to rot and insect infestation. Make sure that proper flashing is specified and detailed on drawings. Do not rely fully on caulks / sealants. As noted at the beginning of **Division 06**, if attaching metal flashing / aluminum **break metal** wrap (coil stock), such can corrode when attached directly to ACQ Pressure Treated (PT) wood. Priming / painting the wood, or covering it with felt (tar) paper first before wrapping it with the aluminum, are a few options that are done.

Also, it should be noted that break metal nailed to wood trim around windows and doors can be an attractive option, promoting low, or no-maintenance opportunity. This unfortunately can trap moisture and allow wood rot on the vary wood that was trying to be protected. *See CSI Section **08 80 00 - Glazing,** CSI Section **08 52 00 - Wood Windows,** and CSI Section **10 81 00 - Pest Control Devices** (flashing), for related information.*

The required locations for exterior flashing, per the **International Building Code** (IBC), *Section **1405.4,*** include, but not limited to:

- At all roof penetrations, especially where a vertical surface (wall, chimney, vent, etc.) intersect a roof.
- Around window and door installations.

Similar information can be found in the **International Residential Code** (IRC), *Section **R317 - Protection of Wood and Wood Based Products Against Decay***, and *Section **R318 - Protection Against Subterranean Termites***

Use this website for *PremiumACCESS or* free *PublicACCESS* viewing of ***I-Codes** and **ICC Standards***:
https://codes.iccsafe.org/

Additional code references that relate to Division 07 are International Building Code (IBC), ***Chapter 13**, Energy Efficiency*, ***Chapter 14**, Exterior Walls,* and IRC, ***Chapter 11**, Energy Efficiency* as well as others listed within the following sections.

The following website is valuable to see which States have adopted which Codes. **http://www.cmdgroup.com/building-codes/**

Before we go any further, it is important to emphasize the assembly of products, including Wall Cladding related flashings. **http://www.copper.org/applications/architecture/arch_dhb/arch-details/wall_cladding/**

*See also, CSI Section **10 81 00 - Pest Control Devices** for **flashing** related information*

07 05 10 - Rainscreen Drainage Mat

- **All About Rainscreens . . .**
  http://www.greenbuildingadvisor.com/blogs/dept/musings/all-about-rainscreens
  (if problems, search ' *All about Rainscreens - Green Building Advisor* ')

- **Archovations Inc.**
  701 Second Street,
  Hudson WI 54016
  Phone: 715.381.5773
  http://archovations.com/rainscreen-mat/

  "CavClear® Rainscreen Drainage Mat *is designed to be installed behind manufactured or thin-set stone, stucco, cedar and lap siding applications to manage moisture and moisture vapor."*

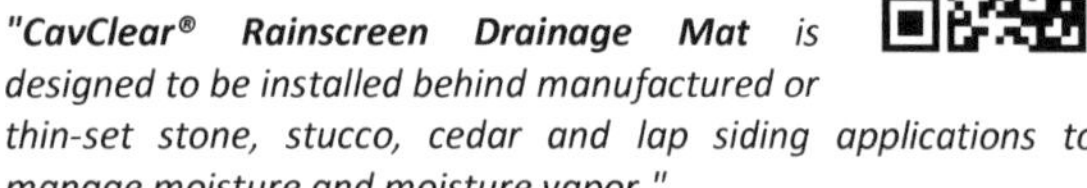

- **Stuc-o-Flex International, Inc.**
  Corporate office location -
  Stuc-O-Flex International, Inc.
  17639 NE 67th Court
  Redmond, WA 98052

  Toll Free: 800.305.1045
  Phone: 425. 885.5085
  Fax: 425.869.0107

  Eastern Division
  1605 Prosser Road
  Knoxville, TN 37914

  http://www.stucoflex.com/rainscreen_drainage_mats.html

  The Stuco-Flex "***WaterWay Rainscreen Drainage Mats*** *consists of a polypropylene core of fused, entangled filaments in varying thicknesses from a nominal 1/8 inch to 3/4 inch, depending on desired cavity space and cladding selection. The core is bonded to a Moisture Resistive Filter Fabric that functions like an additional layer of Weather Resistant Barrier. WaterWay Rainscreen is designed for Stucco, EIFS, Manufactured & Natural Stone, Fiber-Cement, Brick, Lap and Cedar Sidings and other Wall Cladding Systems."*

07 10 00 - Dampproofing and Waterproofing

- **Controlling Unwanted Moisture in Historic Buildings**
  National Park Service
  US Department of the Interior
  Technical Preservation Services
  Preservation Brief 39

  https://www.nps.gov/tps/how-to-preserve/briefs.htm

- ## Laticrete

One LATICRETE Park North
Bethany, CT 06524-3423
Toll Free: 800.243.4788
Phone: 203.393.0010

https://laticrete.com/en/our-products/tile-and-stone-installation/waterproofing

The GreenGuard Certified **Laticrete** ***"9235 Waterproofing Membrane*** *is a thin, load-bearing waterproofing designed specifically for the special requirements of ceramic tile, stone and brick installations. A self-curing liquid rubber polymer and a reinforcing fabric are quickly applied to form a flexible, seamless waterproofing membrane that bonds to a wide variety of substrates."*

- ## Rust-Oleum® - Zinsser

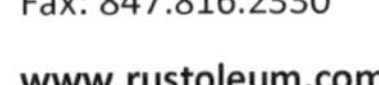

11 Hawthorn Pkwy.
Vernon Hills, IL 60061
Toll Free: 800.323.3584
Phone: 847.367.7700
Fax: 847.816.2330

www.rustoleum.com

"Stop water in its tracks with ***Rust-Oleum Zinsser WATERTITE®-LX Mold & Mildew-Proof Waterproofing Paint****. This low-odor, latex-based formula resists up to 20 PSI of water pressure when applied to concrete or masonry indoors or out, above- or below-grade."*

- ## Mulseal - Tremco Inc.

3735 Green Road
Beachwood, OH 44122
Toll Free: 800.852.9068
Phone: 216.292.5000
Fax: 216.292.5167

http://www.tremcobarriersolutions.com/

Tremco, Inc. offers *"the highest performing waterproofing membranes in the business"*, including the guaranteed Tuff-N-Dri System. Mulseal - Tremco damp proofing was

It is understood that **Mulseal - Tremco damp proofing** is the product used in the ***"All-American Home"*** described at the front of the book, built recently by **Anders Lewendal** and crew - in Bozeman, MT.

For additional **07 10 00** related LEED products, see:
http://www.arcat.com/divs/sec/sec071000.shtml

07 13 00 - Sheet Waterproofing

- ## W. R. Meadows, Inc.

3111 E. 17th Street
Kansas City, MO 64127-2647
Phone: 816-283-3240

https://www.wrmeadows.com/concrete-waterproofing/sheet/

W. R. Meadows, Inc. provides sheet waterproofing membrane applications for Bridges; Parking Decks; Plaza Decks; Structural Slabs; Vehicular, utility and Pedestrian tunnel projects; Hospitals; Schools; Office Construction, and Commercial projects.

For additional **07 13 00** related LEED products, see:
http://www.arcat.com/divs/sec/sec071300.html

07 14 00 - Fluid-Applied Waterproofing

As a result of tighter buildings, for maximum energy-savings, Fluid-Applied systems that serve as both air and waterproofing membranes are also available. Such sprayed on air and moisture resistant barriers are critical to eliminate the air infiltration that ultimately condenses within the walls. This condensation will cause mold and mildew within the wall cavity.

- **PROSOCO**
  3741 Greenway Circle
  Lawrence, KS 66046
  Toll Free: 800.255.4255

Fax: 785.830.9797

http://www.prosoco.com/

"Controlling the leakage of air and water is fundamental to designing an energy efficient building. Performance tested and proven to withstand extreme weather conditions, ***R-Guard*** *protected buildings (by* ***PROSOCO****) are more durable, resilient and sustainable."*

See related video:
https://www.youtube.com/watch?v=x6PHqOXaYIM 3:37

- **Perm-A-Barrier®**
  GCP Applied Technologies, Inc.
  62 Whittermore Ave.
  Cambridge, MA 02140
  Phone: 617-876-1400

https://gcpat.com/en/solutions/products/perm-a-barrier-air-barrier-system/perm-a-barrier-liquid

*"****Perm-A-Barrier® Liquid*** *is a two component, synthetic rubber, cold-vulcanized, fluid applied air and vapor barrier membrane. It cures to form a resilient, monolithic, fully bonded elastomeric sheet, protecting above grade wall assemblies against the damaging effects of air, vapor and water ingress.* ***Perm-A-Barrier® Liquid*** *provides an effective barrier against air infiltration and exfiltration while minimizing the associated energy loss and condensation problems and its versatility makes it easy to use in detailed areas."*

For additional **07 14 00** related LEED products, see:
http://www.arcat.com/divs/sec/sec071400.shtml

07 20 00 - Thermal Protection

Thermal Protection is key to the occupants comfort and providing longevity to the materials and components within the buildings' **thermal envelope**. resistance to heat flow (known as a materials' **R-Value**), is a valuable concept.

All materials typically have the ability to resist or conduct heat. Some are better than the other, depending on what is intended. As someone researches materials, R-Values typically apply to roof, wall and floor materials. So, with these materials, and as we consider what climate the building might be in, or the yearly season, the greater the R-Value, the better the Roof, Wall and/or Floor system is able to resist heat from either entering or leaving the building. As an example, a material or combination of wall materials that add up to R-11 would be less effective to restrict heat flow than a those that add up to R-29.

Glazing on the other hand, (the technical name for glass) is typically measured by its **U-Factor**. Keep in mind that glass is simply a 'glorified hole in the wall', so no matter how energy efficient the glass is, it still is, well . . . *a 'glorified hole in the wall'!* Yet, the smaller the U-Factor, the better it is able to reduce the heat from leaving the building. With that in mind, one example could be U=.35 and another could be a smaller value of U=.18.

Since an R-Value and U-Factor are reciprocals of each other, one can find the equivalent so to compare one to the other. In other words, 1/R = U. Or, 1/U = R. So, with the above U= .35, the R-Value equivalent would be R= 1/.35 or *only* R-2.86. Then, with the above U= .18, the R-Value equivalent would be R= 1/.18 or *somewhat better* at R-5.56, yet still not good compared to the possible R-29 wall system.

Ok? Now, let's tie this into minimum code requirements. See the 2009 International Energy Conservation Code (IECC) *Section **303.1*** to understand what the minimum U-Factor for **Fenestration** (the arrangement of windows and doors on a building) should be.

These requirements can be compared to what is found in the 2015 IECC.

Use this website for *PremiumACCESS or* free *PublicACCESS* viewing of ***I-Codes*** *and* ***ICC Standards***:
https://codes.iccsafe.org/

For more specifics on Fenestration and how the NFRC is a leader in Energy Performance information, see the **National Fenestration Rating Council (NFRC)**.
http://www.nfrc.org/

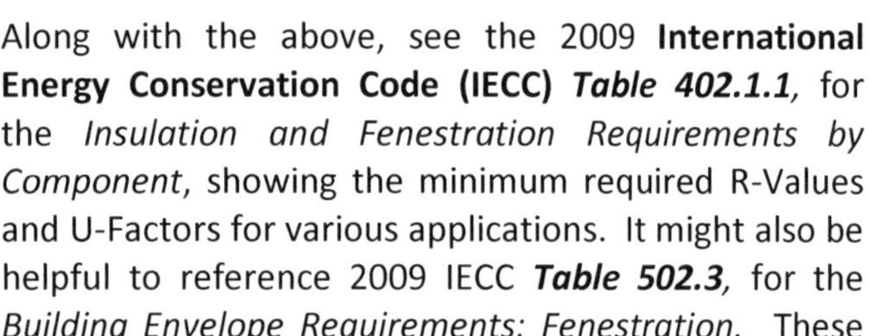

Along with the above, see the 2009 **International Energy Conservation Code (IECC)** ***Table 402.1.1***, for the *Insulation and Fenestration Requirements by Component*, showing the minimum required R-Values and U-Factors for various applications. It might also be helpful to reference 2009 IECC ***Table 502.3***, for the *Building Envelope Requirements: Fenestration.* These requirements can be compared to what is found in the 2015 IECC.

The International Residential Code (IRC) references that may apply include:
- **Chapter 11** - *Energy Efficiency*
- **Section 806** - *Roof Ventilation*
- **Section 807** - *Attic Access* (requirements)

See also ***Division 08 – Openings****, for windows, doors and other 'openings' in a wall or roof.*

As we know, not only is it important to insulate, but also consider: *How the products were made?; How will the products perform over time?; Does the product degrade in any way over time?; Is the product flammable and/or cause noxious smoke?;* And, for whatever the reason, *can the materials be recycled after their use?*

With all the above being considered, let's take a look at various insulating materials.

Insulation Comparisons by the Insulation Institute:

- **http://insulationinstitute.org/im-a-homeowner/about-insulation/insulation-types-comparing-insulation-options/**

Insulation Comparisons by US Department of Energy:

(includes R-Values)

- **http://energy.gov/energysaver/insulation-materials**

Insulation Comparisons by Sustainable Sources

- **http://insulation.sustainablesources.com/**

This resource provides guidelines on the properties of various insulation including the typical products mentioned in the following sections, but also the not so common **Cementitious Foam** and **Perlite**, Cementitious Foam is fire proof, insect proof, and non-toxic. Perlite, a natural extracted material, is used to create insulating light weight concrete slabs and is poured into CMU cores as an insulation. See website for additional information.

Fire Testing Insulation Materials

Comparable Flame / Smoke Tests by Dr. Energy Saver:

- **https://www.youtube.com/watch?v=8NC79e0oztM** 16:21

This is a valuable information. Not necessarily to ASTM standards, but each insulation material is shown how it reacts to flame. When some insulation is shown that it is flame resistant and not giving off any smoke, why would anyone any other

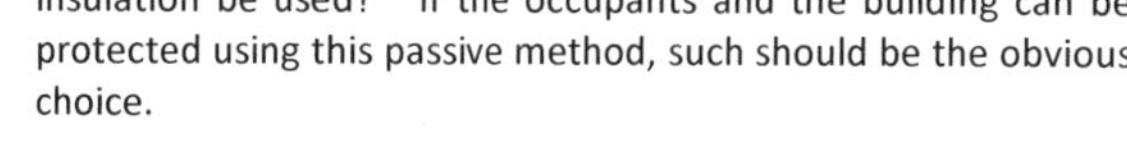

insulation be used? If the occupants and the building can be protected using this passive method, such should be the obvious choice.

- To add to the Fire Testing discussion, see also:

  http://www2.buildinggreen.com/blogs/greensulate-fungus-based-insulation-material-thats-grown-rather-manufactured

- The International Residential Code (IRC) references that may apply include:
  - *Section* ***R316 - Foam Plastics***
  - *Section* ***R302.9.1*** *through* ***R302.9.4 - Flame Spread***

  Use website to access the *I-Codes, PremiumACCESS or the Free PublicACCESS*:
  https://codes.iccsafe.org/

- As listed in *CSI Section* ***06 12 16 - Stressed Skin Panels*** (Structural Insulated Panels). It is noted that the 'Greensulate' mycellium based insulation will not burn, as compared to polystyrene insulation.

Energy Efficiency For Existing Structures

- **Improving Energy Efficiency of Historic Buildings**

  National Park Service
  US Department of the Interior
  Technical Preservation Services
  Preservation Brief 3

  https://www.nps.gov/tps/how-to-preserve/briefs.htm

07 21 13 - Board Insulation

- **EPS Industry Alliance**

  1298 Cronson Boulevard, Suite 201
  Crofton, MD 21114
  Phone: 800.607.3772
  Fax: 410.451.8343

  http://epsindustry.org/

  Expanded Polystyrene Insulation (EPS) sometimes is generally known as Styrofoam™, a trademark of Dow Chemical. It is expanded white beaded, closed-cell insulation that is formed into sheets or other shapes and *"provides minimal water absorption and low vapor permanence."* **It typically has an R-Value of 4.17/inch per 1.0lb density** measured at 40 degrees (F). This R-Value increases with greater density. But, the R-Value is less at warmer temperatures, and the R-Value lessens slightly through age.

  See the above website for a list of manufacturers and other related information. It is understood that Expanded Polystyrene does not contain CFCs or HCFCs.

  EPS can be recycled, but in many ways, at least currently, it is cost preventative. See this website for more information:
  https://www.tucsonaz.gov/es/announcement/styrofoam

- **Extruded Polystyrene Association (XPSA)**

  750 National Press Building
  529 14th Street, NW
  Washington, DC 20045
  Phone: 202.591.2466

  http://www.xpsa.com/

  Extruded Polystyrene (XPS) "*is manufactured in a proprietary process that melt plastic resin and additives into a molten material that is extruded through a die where it expands and cools into a uniform closed-cell rigid foam insulation board with no voids or pathways for moisture to enter."* This insulation is commonly used behind a masonry veneer, used to insulate a foundation wall or below a concrete slab. **It typically has an R-Value of 5.0/inch.** This R-Value is less at warmer temperatures, and the R-Value lessens slightly through age. XPS can be recycled, but recycling centers are minimal. It is understood that CFCs or HCFCs are found in extruded polystyrene foam boards.

- ## FoamGlas® Insulation

Pittsburgh Corning Corp. Headquarters
800 Presque Isle Drive
Pittsburgh, PA 15239-2700
Toll Free: 800.359.8433
Phone: 724.327.6100
Fax: 724.325.9704

http://int.foamglas.com/en/building/applications/

Pittsburgh Corning Corp. manufactures *"High performance, long lasting, durable and environmentally friendly insulation for the entire building envelope"*, and pipe insulating applications.

- ## Mineral Wool Insulation

- RainBarrier® Continuous Insulation

Thermafiber, Inc. / Owens-Corning
3711 Mill Street
Wabash, Indiana 46992
Phone: 260.563.2111

http://www.thermafiber.com/products/commercial/rainbarrier-continuous-insulation/

MONOBOARD®DD / TOPROCK® DD / CAVITYROCK ®DD

"Thermafiber products are available with a minimum recycled content of 70%. These percentage recycled contents are based on pre-consumer (post-industrial) content with 0% post-consumer content. The EPA's Comprehensive Procurement Guidelines for building insulation has recommended the level of recovered materials (recycled) for rock wool insulation be 75%."

- Roxul, Inc.

USA Manufacturing Facility
4594 Cayce Rd.
Byhalia, MS 38611
Phone: 662.851.3348

http://www.roxul.com/about/

"The ROCKWOOL Group is the world's leading manufacturer of stone wool insulation. We offer a full range of high-performing and sustainable insulation products for the construction industry.

Our insulation products are based on innovative stone wool technology and today we are the world's leading manufacturer of stone wool insulation. The ROCKWOOL group also includes a range of sister companies delivering specialist solutions for the horticultural, marine and offshore sectors."

- ## Polyisocyanurate Insulation Manufacturers Association (PIMA)

529 14th Street NW, Suite 750
Washington, DC 20045

http://www.polyiso.org/?

See the above website for a list of members, (affiliates, manufacturers, suppliers) and other related information.

Polyisocyanurate (PIR, Polyiso or ISO) is a closed-cell foam insulation typically sandwiched between two facing layers. It commonly is used as a roof insulation for low-slope roofs and a wall insulation behind a masonry veneer. **It typically has an R-Value of 6.0/inch** and a reported **5.6/inch** as a **Long Term Thermal Resistance value (LTTR)**. This R-Value is less at warmer temperatures, and the R-Value lessens slightly through age.

https://www.carlislesyntec.com/view.aspx?mode=media&contentID=4558

Board Insulation: *Some additional information*

- ***Avoid Polystyrene Insulation***
  http://www2.buildinggreen.com/blogs/avoid-polystyrene-insulation

- ***Discovery Channel How It's Made - EPS Products -***
  https://www.youtube.com/watch?v=XXMque-pLhA 5:20

- ***Recycled-Content Extruded Polystyrene Insulation -***
  https://www.buildinggreen.com/news-analysis/recycled-content-extruded-polystyrene-insulation

- ***Foam Recycling Centers -***
  http://epsrecycling.org/

  http://www.foamfacts.com/recycling/

  http://www.homeforfoam.com/?gclid=CLn_maq9tssCFdgWgQod_rwIPw

- ***Green Insulation Group -***
  10 Pullman Street
  Worcester, MA 01601
  Phone:617.975.3300

  http://www.greeninsulationgroup.com/

  Green Insulation Group is a reseller of previously used EPS, ISO, and XPS insulation.

- ***Stone Wool Insulation: How it's Made -***
  https://www.youtube.com/watch?v=t6FWPTZjwLo 5:00

- ***PolyIso -*** (Polyisocyanurate)
  http://www.greenbuildingadvisor.com/articles/dept/musings/cold-weather-performance-polyisocyanurate

- **Neopor**® (EPS with graphite) -
  http://product-finder.basf.com/group/corporate/product-finder/en/brand/NEOPOR

07 21 16 - Blanket Insulation

Traditional **blanket insulation** is commonly made from fiberglass, yet recycled denim (cotton) and sheep wool insulation are also available. The insulation is typically packaged into rolls for attic floor insulation. The lengths vary depending on the thickness (R-Value) and the widths are designed to fit between framing members that are commonly spaced at 16" or 24"o.c. , apart from each other. The same insulation is also made into shorter sections known as **Batt Insulation**. These are commonly used to fit between the wall stud framing members. Like a sleeping bag, the fiberglass blanket/batt should be "fluffed", (not compressed), for maximum R-Value. As with rigid foam insulation, the color of fiberglass insulation is per the manufacturer.

- **Applegate Insulation**
  Cotton Armor Denim Insulation
  1000 Highview Drive
  Webberville, MI 48892
  Toll-Free: 855.626.8866
  Telephone: 517.521.3545
  Fax: 517.521.3597

  http://www.applegatecottonarmor.com/index.php#.VuHt4l-cGe8 with 1:05 video

- **Black Mountain USA Insulation**
  Sheep Wool Insulation
  110 E. Main Street
  Adamstown, PA 19501
  Phone: 727.366.1368

  http://blackmountaininsulationusa.com/

- **Bonded Logic**
  Ultra Touch Cotton/Denim Insulation
  24053 Arizona Ave., Ste. 151
  Chandler, AZ 85248
  Phone: 480.812.9114
  Fax: 480.812.9633

  www.bondedlogic.com

- **CertainTeed Corporation**
  Fiberglass Insulation
  20 Moores Road
  Malvern, PA 19355
  Phone: 610.893.6000

  http://www.certainteed.com/products/insulation/fiber-glass-insulation

- **Johns Manville**
  Fiberglass Insulation
  600 Jaycee Drive
  Hazle Township, PA 18202
  Phone: 570.455.5340

  http://www.jm.com/en/home-insulation/products/fiber-glass/

- **Owens Corning**
  Fiberglass Insulation
  1Corning Pkwy.
  Toledo, OH 43659
  Toll Free: 800.438.7465
  Phone: 419.248.8000

  https://www.owenscorning.com/

- **Roxul, Inc.**
  USA Manufacturing Facility
  4594 Cayce Rd.
  Byhalia, MS 38611
  Phone: 662.851.3348

  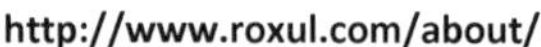

  http://www.roxul.com/about/

  "The ROCKWOOL Group is the world's leading manufacturer of stone wool insulation. We offer a full range of high-performing and sustainable insulation products for the construction industry.

  Our insulation products are based on innovative stone wool technology and today we are the world's leading manufacturer of stone wool insulation. The ROCKWOOL group also includes a range of sister companies delivering specialist solutions for the horticultural, marine and offshore sectors."

Blanket/Batt Insulation: *Some additional information*

- **How It's Made - Fiberglass Insulation**
  https://www.youtube.com/watch?v=UfwlvOh vntw 5:00

- **Fiberglass Insulation: History - Hazards - Alternatives**
  https://www.nachi.org/fiberglass-insulation-history-hazards-alternatives.htm

- **Problems installing cotton insulation**
  http://www2.buildinggreen.com/blogs/problems-installing-cotton-insulation

07 21 26 - Blown Insulation

Blown Insulation historically was limited to cellulose. In recent years fiberglass is also being manufactured for such applications.

- **Applegate Insulation**
  Cellulose Insulation
  1000 Highview Drive
  Webberville, MI 48892
  Toll-Free: 855.626.8866
  Phone: 517.521.3545
  Fax: 517.521.3597

  http://www.applegateweb.com/cellulose-insulation/Cellulose-Insulation-Contractor.shtml

*"**Applegate Cellulose Insulation** contains up to 85% recycled paper, and is up to 50% more effective than other insulations! It's green from start to finish!"*

- ## Blown-In-Blanket System (BIBS)

14100 E. 35th Place, Ste. 104
Aurora, CO 80011
Phone: 800-525-8992
Fax: 303-733-0414

http://www.bibs.com/overview/

*"The **original blow-in blanket system** insulation concept was introduced in the 1950s with a simple goal – to create a system with higher R-values, increased fire resistance, was chemical free, did not settle, provided a consistent application of insulation, and increased energy efficiency. These key performance goals have not changed in almost 50 years and the BIBS system continues to be a leader in building insulation systems today."*

- ## CertainTeed Corporation

TrueComfort Blown-In Fiberglass Insulation
20 Moores Road
Malvern, PA 19355
Phone: 610.893.6000

http://www.certainteed.com/products/insulation/fiber-glass-insulation/blowing-insulation/317372

*"The **TrueComfort** system comprises fiber glass blown-in insulation and the blowing machine used to install it. The machine is portable and simple to operate, and the insulation is super-expanding, requiring fewer bags than other competitive products to achieve the same R-Value"*

- ## Green Fiber

Cellulose Insulation
2500 Distribution St., Suite 200
Charlotte, NC 28203
Toll Free: 800.228.0034
Phone: 704.379.0640

http://www.greenfiber.com/

*"**GreenFiber** is the largest manufacturer of cellulose fiber insulation in the USA". "(The cellulose insulating) products contain a minimum of 85% recycled content . . . (that) contain a mix of pre- and post consumer recycled materials (with) 39% Post minimum consumer material".* **GreenFiber** is an Underwriters Laboratory (UL) Fire Rated product.

- ## Nu-Wool Co., Inc.

Cellulose Insulation
2472 Port Sheldon Road
Jenison, MI 49428
Phone: 616.669.0100

http://www.nuwool.com/

*"Established in 1949, **Nu-Wool Co.,** Inc. manufactures environmentally friendly cellulose insulation materials. Nu-Wool Premium Cellulose Insulation is a superior product designed for use in walls, attics and floors of new and existing residential and commercial structures. Made from recycled paper, Nu-Wool*

Premium Cellulose Insulation has superior thermal and air-infiltration properties, creates quiet and draft-free buildings, and saves up to 40% on energy bills."*

- **Owens Corning**
  AttiCat® Expanding Blown-In Fiberglass Insulation
  1Corning Pkwy.
  Toledo, OH 43659
  Toll Free: 800.438.7465
  Phone: 419.248.8000

  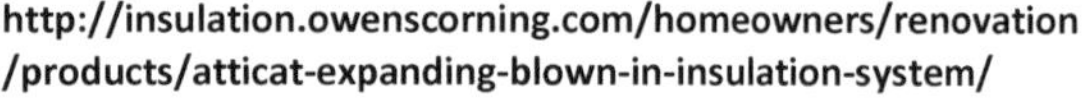

  http://insulation.owenscorning.com/homeowners/renovation/products/atticat-expanding-blown-in-insulation-system/

  *"**AttiCat Blown-In PINK Fiberglas Insulation** products are thermal tested and labeled in strict accordance with current ASTM loosefill standards. Noncorrosive - Doesn't require fire-retardant chemicals which can leach out, promoting corrosion of pipes, electrical equipment or structural metal attachments. Noncombustible - Meets all model building codes for noncombustibility. Won't absorb moisture - Moisture reduces insulation effectiveness."*

Blown Insulation: *Some additional information*

- **How It's Made - Cellulose Insulation**
  https://www.youtube.com/watch?v=feCb8Tzu8SE 5:00

- **National Fiber - Professional resource**
  http://www.nationalfiber.com/genInterest.htm

- **Cellulose insulation -**
  http://www.cellulose.org/userdocs/SpecialReport01-TheBurningQuestion.pdf

07 21 29 - Spray Foam Insulation

Spray Foam Insulation typically has two types; **Open-Cell** and **Closed-Cell**. By its nature, Open-Cell Spray Foam can compress easily and makes a good acoustic insulation barrier for an interior, but because of the open-cells, moisture can enter. Closed-Cell insulation is denser and that makes a good air and moisture barrier for exterior walls, and elsewhere.

Spray Foam typically is made of Petroleum based Polyurethane (SPF). If it is Soy Based, it unfortunately it is said to have approx. 10-15% Soy, and the balance are the other standard chemicals.

- **Spray Polyurethane Foam Alliance (SPFA)**

3927 Old Lee Hwy. #101B
Fairfax, VA 22030
Toll Free: 800.523.6154
Fax: 703.222.5816

http://www.sprayfoam.org/

"Founded in 1987 originally as the Polyurethane Foam Contractors Division, the Spray Polyurethane Foam Alliance (SPFA) is the collective voice, along with the educational and technical resource, for the spray polyurethane foam industry." It is understood that CFCs or HCFCs are found in polyurethane blow-in insulation.

- **SprayFoam.com / SprayFoam Magazine**
  1150 US Route 1 South
  Suite 305
  Jupiter, FL 33477
  Phone: 561-768-9793

  http://www.sprayfoam.com/

  SprayFoam.com and **SprayFoam Magazine** offer resources to find SPF contractors, suppliers and other related information.

Spray Foam Insulation: ***Some additional information***

- **How Can Something So Toxic be Considered Green?**
  http://www.treehugger.com/green-architecture/chemicals-spray-polyurethane-foam-how-can-something-so-toxic-be-considered-green.html

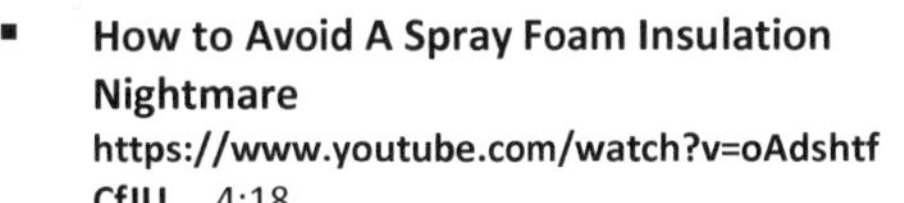

- **How to Avoid A Spray Foam Insulation Nightmare**
  https://www.youtube.com/watch?v=oAdshtfCfJU 4:18

- **Makers of Spray-Foam Insulation Named in Lawsuits**
  http://www.finehomebuilding.com/2013/04/25/makers-of-spray-foam-insulation-named-in-lawsuits

- **Soy Based Foam Insulation** *(typically up to 15% max. Soy)*
  http://www.thegreenestdollar.com/2009/02/soy-based-foam-insulation-what-it-is-and-why-you-should-use-it/

- **Soy, Monsanto, Roundup, and GMO's**
  http://www.motherjones.com/tom-philpott/2014/04/superweeds-arent-only-trouble-gmo-soy

- **US Department of Energy**
  http://energy.gov/energysaver/insulation-materials

- **Touch n Seal Window Insulating Foam**
  http://www.touch-n-seal.com/
  As used at the ***"All-American Home"*** *by* **Anders Lewendal**

- **White Lighting Caulking and Sealants -**
  http://wlcaulk.com/
  As used at the ***"All-American Home"*** ***by*** **Anders Lewendal**

- **Reflectix Sill Sealer -**
  http://www.reflectixinc.com/basepage.asp?PageIndex=690
  As used at the ***"All-American Home"*** ***by*** **Anders Lewendal**

Alternative to Foam Insulation:

- **Greensulate - Ecovative Mycelium Mushroom Insulation**
  Could this C_2C product replace foam insulation?
  http://www.energyvanguard.com/blog-building-science-HERS-BPI/bid/65314/Is-Mushroom-Insulation-the-World-s-Greenest-Insulating-Material

  Ecovative Design
  Eco-HQ
  70 Cohoes Avenue, Suite 103
  Green Island, NY 12183

  Eco-East
  79 102nd St
  Troy, NY 12180

  http://www.ecovativedesign.com/

Ecovative was the **2nd place winner** to the first Cradle to Cradle Product Innovation Challenge. The uses have so far included packaging and SIPs panel alternatives.

http://inhabitat.com/meet-the-winners-of-the-first-cradle-to-cradle-product-innovation-challenge/

07 22 17 - Insulation Baffles (Attic Venting)

There are various insulation baffle type products and eave vent products available. In this case, the specific vent and baffle product is not as critical as the importance to provide such. And, the insulation baffle could be site built. The importance is the air flow should be unobstructed, from the eave vents to the unconditioned attic space. The more vents at the eave or overhang the better! The insulation baffles should prevent the insulation from touching the underside of the roof sheathing.

- **Site-Built Ventilation Roof Baffles**
  http://www.greenbuildingadvisor.com/articles/dept/musings/site-built-ventilation-baffles-roofs

- **"To Vent or Not to Vent"**

https://www.youtube.com/watch?v=Ld8pzIu45F8&nohtml5=False 53:09

Thanks to **Joe Lstiburek** at **Building Science Corporation**, he communicates the importance of a insulation, air-tightness at the ceiling directly below an attic, and venting the steep sloped roof.

07 22 18 - Radiant Barriers

- **Reflective Insulation Manufacturers Assoc., (RIMA) International**
  PO Box 4110
  Olathe, KS 66063
  Toll Free: 800.279.4123
  Fax: 479.657.6749

  http://www.rimainternational.org/

  "RIMA-I is a non-profit industry trade association that exists to propagate the knowledge, use and benefits of reflective insulation, radiant barriers and interior radiation control coatings (IRCCs)."

Radiant Barriers: *Some additional information*

- http://energy.gov/energysaver/radiant-barriers
- http://www.rimainternational.org/index.php/myths/
- http://www.woodbywy.com/osb/weyerhaeuser-radiant-barrier-sheathing-rbs/
- http://www.bondedlogic.com/ultratouch-radiant-barrier/

For a selection of **07 22 18** related LEED products, see:
http://www.arcat.com/divs/sec/sec072218.shtml

07 24 00 - Exterior Insulation and Finish Systems (EIFS)

- **Dryvit**
  One Energy Way
  West Warwick, RI 02893-0914
  Toll Free: 800.556.7752
  Phone: 401.822.4100
  Fax: 401.822.1980

  http://www.dryvit.com/

  Dryvit offers the *"pressure equalized rainscreen"* **Infinity® System** and other **Outsulation®** Exterior Insulation Finish System options

http://www.dryvit.com/fileshare/doc/us/description/ds224_B.pdf

- **Stuc-o-Flex, International, Inc.**
  17639 N.E. 67th Court
  Redmond, WA 98052
  Toll Free: 800.305.5085
  Phone: 425.885.5085
  Fax: 425.869.0107

www.stucoflex.com

*"**Stuc-O-Flex International** created America's first Breathable Elastomeric Acrylic Finish back in 1984. More recently we've advanced the science of **vertical wall Rainscreen in drainable Stucco, Stone, EIFS and siding assemblies** to new levels with technologies unprecedented in past years."*

07 25 00 - Weather Barriers

- **Advanced Building Products, Inc.**
  95 Cyro Drive
  Sanford, ME 04073
  Toll Free: 800.252.2306
  Phone: 207.490.2306
  Fax: 207.490.2998

http://www.advancedbuildingproducts.com/

Advanced Building Products, Inc. is the manufacturer of the **Mortar Break®** and **quality asphaltic** and **non-asphaltic copper laminated flashings**, including various **intangled-net** products.

- **DERBIGUM Americas, Inc.**
  Roofing and Membranes
  4800 Blue Pkwy.
  Kansas City, MO 64130
  Toll Free: 800.727.9872
  Phone: 816.921.0221
  Fax: 816.924.1542

http://www.derbigum.us/

Since 1932, Derbigum has offered various building **water proofing** and **insulation** solutions including its **Intensive** and **Extensive** Green Roofing Systems and solar tube applications.

07 31 00 - Steep Slope Roofing (Shingles & Shakes)

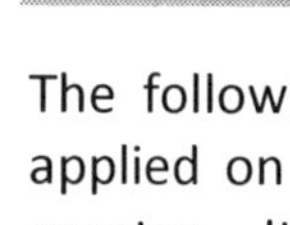

The following sub-sections of 07 31 00 are typically applied on a **Steep Roof** that has a slope of 3/12 or greater. It is wise however to verify manufacturers' warranty information for minimum **roof pitch / roof slope**. See the following sections for the various options.

Roofing warranty information might be misleading. A **lifetime warranty** would truly be wonderful, as it would minimize the tear off and replacement costs, not to mention what goes into the landfill. However, research the manufacturer including possible class action lawsuits, compare products, installers, and understand and compare the fine print.

Asphalt Shingle warranty information:

- http://www.grs-inc.com/?q=node/29

- http://www.bruttellroofing.com/roofing-university/articles/lifetime-asphalt-shingle-warranties-real

07 31 13 - Asphalt Shingles

- **Asphalt Roofing Manufacturers Association**
  Public Information Department
  750 National Press Building
  529 14th Street, NW
  Washington, DC 20045
  Phone: 202.591.2450
  Fax: 202.591.2445

  http://www.asphaltroofing.org/

*"**The Asphalt Roofing Manufacturers Assoc.** (ARMA) is a trade association representing the majority of North America's asphalt roofing manufacturing companies, plus their raw material suppliers."* Contact ARMA for Asphalt Roofing information.

For a selection of **07 31 13** related LEED products, see:
http://www.arcat.com/divs/sec/sec073113.shtml

07 31 26 - Slate Shingles

- **Repair, Replacement, and Maintenance of Historic Slate Roofs**
  National Park Service
  US Department of the Interior
  Technical Preservation Services
  Preservation Brief 29

  https://www.nps.gov/tps/how-to-preserve/briefs.htm

- **National Slate Association**
  P.O. Box 172
  Poultney, VT 05764
  Toll Free: (866) 256-2111

  https://slateassociation.org/

*"The **National Slate Association** was first established in 1922 by producers of slate material. The goal was to promote the use of slate for roofing and structural applications."* Contact the NSA for Distributors, Contractors and other slate roofing information.

For a selection of **07 31 26** related LEED products, see:
http://www.arcat.com/divs/sec/sec073126.shtml

07 31 29 - Wood Shingles and Shakes

Wood is by far one of the most favored of building materials, including its use for siding. In the United States, Pine, Spruce, Fire, Redwood and Cedar are typically used for siding, with Redwood and Cedar being the two that are naturally rot and insect resistant. As I have personally experienced at a residence built in 1956, if cedar is properly attached, using **corrosion resistant nails**, it will not rot during my lifetime. The only minimal locations that suffered rot were where someone used uncoated steel nails for some reason. As these nails rust, this causes the cedar (or any wood) touching these nails to deteriorate over time. Now after personally using ***PC-Petrifier®*** and ***PC-Woody®***, wood restoration products a few years ago, along with a new coat of paint, the original 1956 cedar shake siding looks brand new. This red cedar siding, known as *Pryme-Shakes*, originally from North Tonawanda, NY, appears very similar to the *Cedar Sidewall Shakes* from *Fingerle Lumber Company*. If painted, it is interesting that some people mistake these for asbestos siding, installed years ago!

- ### Cedar Sidewall Shakes
  Fingerle Lumber Company
  Mailing and Sales Address:
  617 S. Fifth Ave.
  Ann Arbor, MI 48104-2905
  Toll Free: 800.365.0700
  Phone: 734.663.3550
  Fax: 734.663.0137

  http://www.fingerlelumber.com/products/cedar/cedar-shakes-shingles

  Along with the cedar shakes and shingles for walls and roofs, **Fingerle Lumber Co.** carriers a full line of construction products

- ### The Repair and Replacement of Historic Wood Shingle Roofs
  National Park Service
  US Department of the Interior
  Technical Preservation Services
  Preservation Brief 19

https://www.nps.gov/tps/how-to-preserve/briefs.htm

For a selection of **07 31 29** related LEED products, see:
http://www.arcat.com/divs/sec/sec073129.shtml

*- See CSI Section **07 46 23 - Wood Siding,** for related products and information, and CSI Section **06 01 00 - Maintenance of Wood, Plastics, and Composites** for wood repair products.*

07 32 13 - Clay Roof Tiles

- **Preservation and Repair of Historic Clay Tile Roofs**
  National Park Service
  US Department of the Interior
  Technical Preservation Services
  Preservation Brief 30

  https://www.nps.gov/tps/how-to-preserve/briefs.htm

- **Tile Roofing Institute**
  2150 N 107th Street, Ste 205
  Seattle, WA 98133
  Phone: 206.209.5300

  http://tileroofing.org/

The **Tile Roofing Institute**, *"founded in 1971, (and) originally named the National Tile Roofing Manufacturers Association (NTRMA) – has produced technical manuals on code language and preferred installation practices within all the major code bodies nationwide.* The Tile Roofing Institute provides membership opportunities and certified contractor / installer contact information.

For a selection of **07 32 13** related LEED products, see:
http://www.arcat.com/divs/sec/sec073213.shtml

07 32 16 - Concrete Roof Tiles

For a selection of **07 31 26** related LEED products, see:
http://www.arcat.com/divs/sec/sec073216.shtml

*- See **CSI Section 07 32 13 - Clay Roof Tiles,** for Tile Roofing Institute related information.*

- Article - *The Basics: Clay and Concrete Roof Tiles,* by Bob Vila, compares items, including costs, weight and durability.
https://www.bobvila.com/articles/tile-roofs/#.WWxH9uSGMdU

07 32 19 - Metal Roof Tiles

- **Metal Roof Alliance**
  12430 Tesson Ferry Road #112
  Saint Louis, MO 63128
  Phone: 314.495.5906

  http://www.metalroofing.com/

  The **Metal Roof Alliance** provides contractor and manufacturer support and contact information for the end user, along with metal roof product options, descriptions and images that are available within the metal roofing industry.

For a selection of **07 32 19** related LEED products, see:
http://www.arcat.com/divs/sec/sec073219.shtml

07 40 00 - Roofing and Siding Panels

- **Metal Construction Association**
  8735 W. Higgins Road, Ste. 300
  Chicago, IL 60631
  Phone: 847.375.4718
  Fax: 847.375.6488

  http://www.metalconstruction.org/

  The **Metal Construction Assoc. (MCA)** provides a manufacturer support and contact information that promotes metal options for **Steep Slope Roof, Low Slope Roof**, and metal wall products that include **Insulated Metal Panels, Metal Composite Panels** and **Single Skin Panels.**

For a selection of **07 40 00** related LEED products, see:
http://www.arcat.com/divs/sec/sec074000.shtml

07 41 43 – Plastic / Rubber Composite Roof Panels

- **EcoStar LLC**
  42 Edgewood Dr.
  Holland, NY 14080
  Toll Free: 800.211.7170

Fax: 888.780.9870

http://www.ecostarllc.com/index.aspx

*"**EcoStar** is the leading manufacturer of sustainable, environmentally friendly steep-slope roofing products. (Their) products are manufactured with recycled rubber and plastics and are highly flexible . . . offering superior performance and durability." Products include **Majestic Slate™** and **Empire Slate™** synthetic slate tiles, **Seneca Shake™** and **Empire Shake™** synthetic shake tiles, a line of designer tiles, and a full line of accessories, including **EnBase™** insulation, **EcoVent™**, Aqua **Guard™**, **Glacier Guard™**, **EcoStar Fasteners™**, and **NOVA Walkway Pads™** to construct a complete roof system.*

For a additional **07 41 43** related LEED products, see:
http://www.arcat.com/divs/sec/sec074133.shtml

07 42 43 - Composite Wall Panels

The nightmarish London fire on June, 14, 2017 - at the **Grenfell Tower**, unfortunately used **Aluminum Composite Material (ACM)** wall panels that had a flammable polyethylene core. Per the below article, Reynobond® PE was used, and is banned in the USA on buildings over 40 ft. (12.192 meters) tall. **http://metro.co.uk/2017/06/16/why-is-cladding-banned-in-the-us-and-germany-used-on-buildings-in-the-uk-6712578/**

From their website listed below, it is understood that Alcoa also manufactures and sells Reynobond® FR, which is a similar product, with a fire-retardant mineral core. Unfortunately, the flammable PE product was specified and installed, possibly saving renovation costs, and potentially helping to spread the recent fire that occurred approximately one year after this early 1970's building was renovated.

It is decisions like this that show the importance of research and choosing one product over the other, while understanding its consequences, including the approximate 80 fire related casualties!

After researching this building a bit further, it is understood that this (and others built similar to it) had only one central stair. With the building renovations, adding a second egress stair may have helped save lives.

Some of the various **Composite Wall Panel** options include:

- **Alucobond®**
  3480 Taylorsville Hwy.
  Statesville, NC 28625
  Toll Free: 800.626.3365

  http://www.alucobondusa.com/

  *"**Alucobond** is the Global Brand Leader. From its creation in 1969, the "Original Aluminum Composite Material," Alucobond has earned its brand leadership through consistent performance and service excellence."*

- **Citadel Architectural Products, Inc.**
  3131-A North Franklin Road
  Indianapolis, IN 46226
  Toll Free: 800.446.8828
  Phone: 317.894.9400
  Fax: 317.894.6333

  http://www.citadelap.com/

  Citadel Architectural Products, Inc. manufacturers a variety of economical and durable composite exterior cladding options that are field assembled or shop fabricated.

- **Reynobond®** **PE** and **Reynobond®** **FR**
  Alcoa Architectural Products
  50 Industrial Boulevard
  Eastman, GA 31023-4129

  https://www.arconic.com/aap/north_america/en/product.asp?cat_id=915&prod_id=1534

  "Reynobond PE is flexible to the core.
  Reynobond PE features a polyethylene core that adds strength and rigidity to the coil-coated aluminum panels. This maximizes its flexibility and formability, while maintaining a light weight for easy installation.

  Reynobond FR adds an extra layer of protection.
  Reynobond FR is manufactured just like Reynobond PE, but with a fire-retardant mineral core that guarantees higher resistance to fire. These panels meet or exceed national model building code requirements without exception."

For a additional **07 42 43** related LEED products, see:
http://www.arcat.com/divs/sec/sec074243.shtml

07 42 63 - Fabricated Wall Panel Assemblies

- **ATAS International, Inc.**
  6612 Snowdrift Road
  Allentown, PA 18106
  Toll Free: 800.468.1441
  Phone: 610.395.8445
  FAX: 610.395.9342

  http://www.atas.com/

ATAS International, Inc. designs and manufacturers sustainable products, in sustainable ways. See their selection of Insulated Metal Wall Panels, along with Metal Roofing, Roof Edge / Coping, Water Control Systems, and Structural Members. It also manufactures and produces *InSpire Wall*, an aluminum passive heating cladding system. See website or Division 23 for additional information.

- **Centria**

  1005 Beaver Grade Road
  Moon Township, PA 15108
  Toll Free: 800.759.7474
  Fax: 412.299.8317

  http://www.centriaperformance.com/products/products.aspx

  *"**CENTRIA** has been leading the way with high performance, highly aesthetic architectural systems and building envelopes for over 100 years. Our comprehensive line of exterior metal building products and services help give a dramatic edge to some of the world's finest commercial, institutional and industrial buildings."*

- **Kal-Wall**

  1111 Candia Road
  Manchester, NH 03109

  https://www.kalwall.com/

  *"**Kalwall** strives to be leaders in the building industry by maximizing energy savings and harvesting free sunlight while providing affordable, productive, healthy environments for all building occupants. All of our daylighting products rely on our innovative, sustainable, lightweight structural sandwich panels that are revolutionary in fenestration technology for their combined light quality, energy efficiency and solar control properties."*

- **Panelite**

  Professional Services:
  315 W 39th St # 807A
  New York, NY 10018
  Phone: 212.947.8292

  http://www.panelite.us/

  Showroom:
  8599 Venice Boulevard
  Los Angeles, CA 90034
  Phone: 323-297-0115

  http://www.panelite.us/projects/exterior-projects-gallery/

  *"**Panelite** develops innovate, energy-efficient materials for Architecture"* that apply to interior and exterior applications. Panelite created the *"translucent honeycomb panel in 1998"*.

- **Stone Panels International, Inc.**

  100 S. Royal Lane
  Coppell, TX 75019
  Phone: 800.328.6275

  http://www.stonepanels.com/

  Stone Panels, Inc., under the name Stone Lite® creates natural stone panels mounted to a honeycomb aluminum backing for

the interior and/or exterior of many various building types. This offers a solid stone alternative *"when dimensional stone is too heavy to be economical."*

Building Envelope: *Some additional information*

- ### National Institute of Building Sciences: WBDG

  1090 Vermont Ave NW # 700
  Washington, DC 20005
  Phone:202.289.7800
  Fax: 202.289.1092

  Building Envelope Design Guide: Panelized Metal Wall Systems -

  https://www.wbdg.org/design/env_wall_panelizedmetal.php

- ### RCI, Inc.

  Vacuum Insulated Panels
  1500 Sunday Drive, Suite 204
  Raleigh, North Carolina 27607
  Toll Free: 800.828.1902
  Phone: 919.859.0742
  Fax: 919.859.1328

  http://www.rci-online.org/

  "RCI, Inc. is an international association of building envelope consultants. Our members specialize in design, investigation, repair and management of roofing, exterior wall, and waterproofing systems."

  RCI, Inc. article addressing ***Vacuum Insulated Panels, (VIPs)***, their future, and their R-values of R-60 / inch - compared to other insulation.

- ### Thermal Vision

  Vacuum Insulated Panels
  83 Stonehenge Drive
  Granville, OH 43023
  Phone: 740-973-3671
  Fax: 740-587-4025

  http://www.thermalvisions.com/Theshhold_page.html

  "Threshold, the next Step Vacuum Insulation Panel - VIP New R50" *[by Thermal Vision]*, can be used for Scientific freezers; Water heaters; Shipping Containers and other related applications as listed at the above website.

- ### Related article about Vacuum Insulated Panels (VIPs)

  Meeting Efficiency Codes Without Compromising Design -

  http://www.constructionspecifier.com/meeting-efficiency-codes-without-compromising-design/

- ### Aspen Aerogels, Inc.

  Headquarters

30 Forbes Road
Northborough, MA 01532
Toll Free: 888.481.5058
Phone: 508.691.1111

Manufacturing Facility
3 Dexter Road
East Providence, RI 02914
Phone: 401.432.2612

http://www.aerogel.com/

Aspen Aerogels, Inc. is the manufacturer of various high performance aerogel insulation that *"offers extremely low thermal conductivity in a thin, lightweight design . . . (that are used worldwide, in a) wide variety of other applications including industrial piping, transportation, outdoor gear, and building construction/renovation."*

- ## Cabot Aerogel

4400 North Point Parkway, Suite 200
Alpharetta, GA 30022
Phone: 678.297.1300
Fax: 678.297.1245

http://www.cabotcorp.com/solutions/products-plus/aerogel

"Aerogel, known as the "world's best insulating solid material," is used to enhance the thermal performance of energy-saving materials and sustainable products for buildings, on- and off-shore". See Cabot Corp. website for products.

- ## NanoPore Insulation, LLC

3271 Spirit Drive SE
Albuquerque, NM 87106
(505) 224-9373 Phone
(505) 224-9376 Fax

http://www.nanopore.com/index.html

*"**NanoPore™** thermal insulation is a porous solid that is prepared by one of several processes which yield both low density and small pores. Its chemical composition is silica, titania and/or carbon in a 3-D, highly branched network of primary particles. NanoPore™ Thermal Insulation can have thermal resistance values as high as R40/inch - 7-8x greater than conventional foam insulation materials. It is (the) nano-scale porosity that gives **NanoPore™** its excellent thermal performance."*

- ## Buy Aerogel.com

The website sells Aspen Aerogel and Cabot Aerogel products. See product information and prices -
http://www.buyaerogel.com/product/spaceloft/

"AEROGEL - THE MOST EFFECTIVE THERMAL INSULATOR ON EARTH" - Aspen Aerogels, Inc.

*- See CSI Section **13 31 00 - Fabric Structures,** for Aerogel related products and information.*

07 44 16 - Porcelain Enameled Faced Panels

- **ICAP-USA / InsulStone**
  250 E. Franklin Rd.
  Meridian, ID 83642
  Phone: 208.895.8557
  Fax: 208.639.6422

  http://www.icap-usa.com/

  ICAP-USA / InsulStone offers patented insulated cladding architectural products that are porcelain, natural stone or cast stone finish, that provide R-values of R-6, R-12 and R-18.

- **TerraCORE Panels**
  2030 Irving Blvd.
  Dallas, TX 75207
  Toll Free: 877.503.2062
  Phone: 214.749.0999
  Fax: 214.742.2804

  http://www.terracorepanels.com/ with 3:07 video

  TerraCORE Panels offers lightweight, environmentally friendly, and affordable luxury porcelain, natural stone and metal wall panels for interior or exterior new and renovation based projects.

For additional **07 44 16** related LEED

products, see:
http://www.arcat.com/divs/sec/sec074416.shtml

07 46 23 - Wood Siding

As also stated previously in *CSI Section* ***07 31 29 - Wood Shingles and Shakes,*** wood is by far one of the most favored of building materials, including its use for siding. In the United States, Pine, Spruce, Fir, Redwood and Cedar are some of the most common woods produced for siding, with Redwood and Cedar being the two that are naturally rot and insect resistant.

- **Bark House**
  534 Oak Avenue
  Spruce Pine, NC 28777
  Phone: 828.765.9010

  http://barkhouse.com/

  Bark House manufacturers the original reclaimed Poplar Bark Shake that is a Cradle-to-Cradle, maintenance-free, timeless option which has stood the test of time and is used to *"adorn and protect the finest homes, lodges, and commercial buildings across every region of the United States and Internationally."*

- **Louisiana Pacific** (LP)
  LP Marketing Center
  One Thomas Circle, NW
  Suite 600
  Washington, DC 20005
  Phone: 888-820-0325
  Fax: 877-523-7192

  https://lpcorp.com/products/siding/lp-smartside-trim-siding/

  Along with various other wood products, **Louisiana Pacific** manufactures a variety of exterior wood options, including **LP Smartside Trim & Siding; LP Canexel Prefinished Siding; and LP Armorstrand Substrate**

- **Wood Haven, Inc.**
  401 Bridge St.
  Perry, KS 66073
  Toll Free: 800.545.8884
  Phone: 785.597.5618

  http://www.rainscreenclip.com/sidings-decks/

  "Traditional wood siding is a superior choice for your building's exterior. Unlike other cladding options, wood siding is 100% natural and durable. When compared to other building material options, wood siding has a very small carbon footprint when responsibly harvested. ***Wood Haven, Inc.*** *is dedicated to ensuring that our wood comes from reliable, quality, and responsible companies."*

For additional **07 46 23** related LEED products, see:
http://www.arcat.com/divs/sec/sec074623.shtml

Wood Siding: ***Some additional information***

- See Section ***07 31 29 - Wood Shingles and Shakes*** *for related products and information for both roof and siding applications.*

- ***T1-11 Siding***
http://www.homeadvisor.com/r/t1-11-siding/#.WW0SEuSGMdU

07 46 35 - Insulated Vinyl Siding

- **Aluminum and Vinyl Siding on Historic Buildings**
  National Park Service
  US Department of the Interior
  Technical Preservation Services
  Preservation Brief 8

  https://www.nps.gov/tps/how-to-preserve/briefs.htm

- **Vinyl Siding Institute, Inc.**
  1201 15th St NW, Suite 220
  Washington, DC 20005
  Phone: 202.587.5100

  https://www.vinylsiding.org/

  The Vinyl Siding Institute provides related information to homeowners, Designers and Installers, including a locator service for Certified Installers and Certified Products.

- **Inspectapedia Vinyl Siding - Related information:**
  http://inspectapedia.com/exterior/Vinyl_Siding_Guide.php

For additional **07 46 23** related LEED products, see:
http://www.arcat.com/divs/sec/sec074600.shtml

07 46 46 - Fiber-Cement Siding

- **Portland Cement Association** (PCA)
  5420 Old Orchard Road
  Skokie, Illinois 60077-1083
  Toll Free: 847.966.6200

  Fax: 847.966.9781

  http://www.cement.org/structures/buildings-structures/concrete-homes/products/fiber-cement-siding

  "Although fiber cement siding commonly has the look of wood siding, it can be manufactured in a variety of profiles. In all cases, it is a cost effective and low maintenance finish that is resistant to termites and fire, will not rot, buckle or warp, and holds paint for several years longer than conventional wood siding." The **Portland Cement Assoc. (PCA)** represents the Fiber-Cement Siding industry.

For a selection of **07 40 00** related LEED products, see:
http://www.arcat.com/divs/sec/sec074646.shtml

07 50 00 - Membrane Roofing

Just as *CSI Section* ***07 31 00*** represented steep slope roofing, (roofing that typically can be applied to a roof slope of 3/12 or greater), *CSI Section* ***07 50 00*** represents **low slope roofing**, (roofing materials that typically are applied to a roof that has a slope commonly less than 3/12). Since there is no such thing as a 'flat roof', all roofs, per code, should have some slope to them as a way to shed water towards **internal roof drains** or **perimeter roof drains**.

The low slope roofing materials that have become very common are single-membrane type roof materials. These single-membrane type roofs include: **Thermoset Membranes** (which commonly are synthetic rubber, such as EPDM); **Thermoplastic Membranes** (which commonly are PVC based, including Thermoplastic polyolefin (TPO) and others); and **Modified Bitumen Membranes** (which combine asphalt with reinforcement materials).

The topic of **cool roofs** (a light-colored, reflective roof) can apply to this CSI section, but also applies to any roof material *"that has been designed to reflect more sunlight and absorb less heat"*.

See: **http://energy.gov/energysaver/cool-roofs**

Yet, are cool roofs green? Based on recent studies, there is a controversy if cool / light-colored / reflective roofs are really beneficial in <u>all</u> climate zones. *"Over the last 10 years, white reflective roofs - widely touted in southern climates for their ability to help lower air-conditioning demand and, thus energy bills, conserve natural resources, combat urban heat-island effect, and reduce global-warming - have been migrating north. Yet, contrary to news articles and politicians promoting "that most northern US Cities are wasting millions of dollars in energy, as well as precious natural resources, by buying into the belief white roofs are energy-efficient regardless of climate zone."* See the complete article:

The Myth of White Roofs in Northern Climates-
https://www.carlislesyntec.com/download.aspx?fileID=6675

With a little more research, the debate continues with this article:

Are Cool Roofs Green? The Answer is Not So Black and White
http://www.epdmroofs.org/attachments/are%20cool%20roofs%20green.pdf

- **EPDM Roofing Association** (ERA)
  529 14th Street, NW, Suite 750
  Washington, DC 20045
  Phone & Fax: 202.591.2474

  http://www.epdmroofs.org/

  Ethylene Propylene Diene Monomer (EPDM) is a synthetic rubber commonly used in low-slope, single-play membrane roofing. The EPDM Roofing Association *"provides technical and research support to the public and the construction industry, and communicates the longstanding attributes, consistency and the value proposition of EPDM rubber membrane roofing materials."* EPDM has shown to have Long Term service, Life Cycle Assessment while also being recyclable, where *"3.25 million* *pounds of reclaimed EPDM membrane has been diverted from landfills".*

- **National Roofing Contractors Assoc. (NRCA)**
  10255 W. Higgins Road, Suite 600
  Rosemont, IL 60018-5607
  Telephone: 847.299.9070
  Fax: 847.299.1183

  http://www.nrca.net/

  ***NRCA** has been the professional voice of its members since 1886. With more than 3,600 members, NRCA is the leading authority in the roofing industry for information, education, technical assistance, advocacy, publications, programs and support that help your business grow."*

For a selection of **07 50 00** related LEED products, see:
http://www.arcat.com/divs/sec/sec075000.shtml

07 55 64 - Green Roof Components

The **roof garden** concept goes back at least to the famous *Hanging Gardens of Babylon,* where buildings were adorned with, and encompassed plant life. Then, fast forward to the late 1970's - 1980's, **Earth Sheltered** residential construction was popular, making the use of the natural thermal qualities of soil, covered with vegetation, yet for some reason. In recent years, it is interesting to see how something from the past has become popular again, but now as an expected trademark of a sustainable building. However, it is important to keep in mind that such green roofs (**living roofs / vegetative roofs**), typically require additional structural considerations to compensate for the added weight offered by the plantings; the water saturated plant media; the root barrier, drainage system; pavers; possible irrigation systems; the live loads of 'x' number of people on the roof; and any other needs or desires. As you know, the weight would differ depending on if the green roof was designed as **extensive** (less growing media with smaller plants) to one that would be **intensive** (greater amounts of growing media with larger plants and trees).

Without anticipating such dead and live loads, a disaster like what happened at City University, Hong Kong is inevitable.
https://www.youtube.com/watch?v=P3hoadjJCyg

Such green roofs are beautiful but also require additional maintenance than a typical membrane roof.

And, what happens when the roof membrane needs replacing after the 20 year warranty?! Can you imagine the replacement costs involved to remove and replace the plants, growing media, drainage system, and other components?! So, obviously when one is proposed it should be considered as a valuable extension of the interior. An outdoor room. A space that otherwise is not possible with a restricted building lot. So, making use of the roof square footage becomes a useful space, thus a financial benefit. With this, the best building location for such would be in an urban environment where on-grade green spaces are almost non-existent and are considered a premium. This is where a buildings' roof deck(s), vegetation and potential trees could be enjoyed from within, at multiple vantage points, while also being easily accessible at different times, for different reasons. The vegetation would offer the beauty, and its natural air cleaning benefit, while being irrigated by the cleansing rain. A 'win-win' for many reasons, for such a location.

- **Carlisle SynTec Systems**
  P.O. Box 7000
  Carlisle, PA 17013
  Toll Free: 800.479.6832

https://www.carlislesyntec.com/template.aspx?page=templat e&category=149

*"**Carlisle's Roof Garden Systems** provide a variety of waterproofing options with a full line of accessories to ensure a high-performance system for virtually any type of application or building type. We offer both traditional, planted-in-place green roof options as well as a diverse line of modular systems."*

- **DERBIGUM Americas, Inc.**
  4800 Blue Pkwy.
  Kansas City, MO 64130
  Toll Free: 800.727.9872
  Phone: 816.921.0221
  Fax: 816.924.1542

http://www.derbigum.us/solutions/green-roofs

Derbigum offers **Intensive** and **Extensive** Green Roofing Systems that include recyclable roofing membranes. See above website for more information.

- **J-Drain**
  JDR Enterprises, Inc.
  292 South Main Street, Ste. 200
  Alpharetta, GA 30004
  Phone 770.442.1461
  Fax 770.664.7951

http://www.j-drain.com/

*"**JDR Enterprises, Inc.** is a leading manufacturer and supplier of greenroof drain components. As the market for green roofing develops, the proper design and understanding of green roof components is imperative. JDR Enterprises, Inc. encourages greenroof technologies and will continue to do its part by providing key components for greenroofs."*

- ## LiveRoof, LLC.

A subsidiary of Hortech, Inc.
Mailing address:
PO Box 533
Spring Lake, MI 49456

http://www.liveroof.com/representatives/

LiveRoof Licensed Growers are regional horticultural experts which oversee the local design, sale, and growing of LiveRoof projects. See above website for a regional grower.

- ## Roof Meadow

7135 Germantown Avenue, 2nd Floor
Philadelphia, PA 19119-1842
Phone: 215.247.8784
Fax: 215.247.4659

www.roofmeadow.com

*"**Roofmeadow** is an award-winning green roof civil engineering and design firm specializing in elegant and inventive designs. Roofmeadow's multidisciplinary staff of landscape architects, civil engineers, horticulturalists and construction specialists".*

- ## XEROFLOR North America

825Third Avenue, 2nd Floor
New York, NY 10022
Phone:866-335-1151

https://xeroflornorthamerica.com/

"Our green roof mats are supplied by local, independent farms located across the continent and our plants are grown with regional horticultural expertise and adapted to the climates where our systems are installed. No green roof system available in North America is easier to install than XeroFlor. And no green roof system in North America offers more performance validation based on independent scientific research than Xero Flor."

07 61 00 - Sheet Metal Roofing

- ## Centria Architectural Systems

1005 Beaver Grade Road
Moon Township, PA 15108-2944
Toll Free: 800.759.7474
Phone: 412.299.8000
Fax: 412.299.8317

http://www.centria.com/Pages/default.aspx

Centria Architectural Systems produces Structural Design Panels for roofs and the Curved Standing Seam Roof systems. Along with this Centria also offers Intercept Modular Metal Panel System and other metal systems for the building envelope.

- **Peterson Aluminium Corp. (Pac-Clad)**
  1005 Tonne Road
  Elk Grove Village, IL 60007
  Toll Free: 800.323.1960
  Phone: 847.228.7150
  Fax: 800.722.7150

  https://www.pac-clad.com/#

  Pac-Clad manufactures metal roofing systems with both concealed and exposed fasteners; metal soffit panels; metal column covers; metal perimeter / roof edge (fascia & coping) systems; air-flow systems; metal wall panels and other metal fabrications.

For additional **07 61 00** related LEED products, see:
http://www.arcat.com/divs/sec/sec076100.shtml

07 70 00 - Roof and Wall Specialties and Accessories

Roof and Wall Specialties and Accessories typically include the following: **Roof Edge Coping; Counter Flashing; Gravel Stops; Gutters & Downspouts; Fascia; Reglets; Roof Expansion Joints; Roof Suppers; Ridge Vents; Relief Vents; Roof Hatches; Snow Guards; Walk Boards & Roof Pavers.**

For a selection of **07 70 00** related LEED products, see:
http://www.arcat.com/divs/sec/sec077000.shtml

07 80 00 - Fire and Smoke Protection

Fire and Smoke Protection typically include the following: **Cementitious and Intumescent Fireproofing; Fire stopping; Smoke Seals; and Smoke Containment Barriers.**

For a selection of **07 80 00** related LEED products, see:
http://www.arcat.com/divs/sec/sec078000.shtml

The **International Buidling Code** (IBC) Sections that could apply, include:
- **Chapter 7** - *Fire and Smoke Protection Features*
- **Chapter 10** - *Means of Egress*
- **Chapter 11** - *Accessibility*

- **Chapter 12** - *Interior Environment*

And, the **International Residential Code** (IRC) Sections that could apply, include:
- **Chapter 3 - Building Planning, including:**
- **Section R302** - *Fire-Resistant Construction*
- **Section R302.7** - *Under-stair Protection*

07 90 00 - Joint Protection

- ### Construction Specialties (C/S)

  6696 State Rt. 405
  Muncy, PA 17756
  Toll Free: 800.233.8493
  Phone: 570.546.4505
  Fax: 570.546.5169

  http://www.c-sgroup.com/

  Construction Specialties manufacturers sustainable options for Floor Joint Covers; Wall & Ceiling Joint Covers; Exterior Specialty Covers; Parking Garage & Stadium Covers; and "*Restofit™ Joint Systems provide refreshing options for worn out and broken down expansion joint covers*"; along various Division 10 products.

- ### Schluter® Systems

  194 Pleasant Ridge Rd.
  Plattsburgh, NY 12901-5841
  Toll Free: 800.472.4588
  Fax: 800.477.9783

  http://www.schluter.com/schluter-us/en_US/

  Schluter® Systems manufactures various Mortar Bed Joint Profiles; Surface Joint Profiles; Perimeter Joint Profiles; and Cove-shaped Profiles all designed for finished floor and wall applications.

For additional **07 90 00** related LEED products, see:
http://www.arcat.com/divs/sec/sec079000.shtml

Thermal & Moisture Protection: *Some additional information*

- ### "The Perfect Wall" . . .

  http://buildingscience.com/documents/insights/bsi-001-the-perfect-wall

  Thanks to **Joe Lstiburek** and the great folks at **Building Science Corporation**, not only does this concept apply to walls, but it also applies to roofs and floor systems too. Take time to review and apply it to your next project!
  (This is a very important topic with Sustainable Design.)

- **Perfect Wall Review - A System To Build An Efficient House**
  https://www.youtube.com/watch?v=hTwq-qUnr9I 4:46

See related **Building Science** information and Building Science Corporation links at the end of this books' ***Introduction*** heading.

*- See CSI Section **13 31 00 - Fabric Structures,** for related products and information for fabric Tensioned Membrane Structures and Air-supported roofs.*

End of Division 07

Space for notes:

DIVISION 08 - Openings

Openings in a wall, sometimes known as ***fenestration,*** (the arrangement of doors & windows), represent by far one of the most appreciated elements of a building. This section includes ***windows***, ***doors, vents, louvers, roof windows*** and ***skylights*** that adorn our buildings with character. This is especially true with those carefully detailed and proportioned elements found on the various historic buildings that are particularly unique to each rural community, town, and city across America.

With such openings, it is important to understand and apply related building codes for **egress, daylighting** and **ventilation** requirements - as it would relate to the new, alterations and addition(s) based projects you might be involved in. Keep in mind that egress is not limited to leaving the building. Depending on the what the building is designed for, in other words, the building **Use and Occupancy Classification**. If the building has multiple uses and/or various tenants, the corridor leading to the public way will need to be fire rated along with the doors and windows (openings) along this corridor.

Obviously, it makes no sense to design and select products without understanding if such meets minimum code requirements. As mentioned in elsewhere in this book, free access to the **Building Codes** is available at: **http://codes.iccsafe.org/index.html**

To reference which codes apply to which States, see: **http://www.cmdgroup.com/building-codes/**

Depending on the application, the **International Building Code** (IBC) chapters that could apply, are:
- **Chapter 3** - *Use and Occupancy Classification*
- **Chapter 4** - *Special & Detailed Reqt's based on Use and Occupancy*
- **Chapter 5** - *General Building Heights and Areas*
- **Chapter 6** - *Types of Construction*
- **Chapter 7** - *Fire and Smoke Protection Features*
- **Chapter 10** - *Means of Egress*
- **Chapter 11** - *Accessibility*
- **Chapter 12** - *Interior Environment*
- **Chapter 13** - *Energy Efficiency*

And, the **International Residential Code** (IRC) Sections that could apply, include:
- **Section R302.6** - *Dwelling/Garage Fire Separation*
- **Section R303** - *Light, Ventilation & Heating*
- **Section R308** - *Glazing*
- **Section R310** - *Emergency Escape & Rescue Openings*
- **Section R311** - *Means of Egress*
- **Section R612** - *Exterior Windows and Doors*

One particular specific requirement that applies to both IBC and IRC is the minimal amount of ventilation and daylighting that is required for **Habitable** (Living, Eating, Sleeping and Cooking) spaces. This gets into the **Indoor Air Quality (IAQ)** and the **Indoor Environmental Quality (IEQ)** topics. Take a look at IBC Section **1203** - *Ventilation*; IBC Section **1205** - *Lighting*; and IRC Section **R303** to learn more.

If the structure has historical significance and/or is Registered as a **National Historic Landmarks Program** (**https://www.nps.gov/nhl/**), or the structure is listed on the **National Register of Historic Places** (**https://www.nps.gov/nr/**), be mindful that such windows and doors commonly can be restored, verses being replaced. And, if replaced, they need to be replaced with 'like kind' materials, profiles and proportions. Many times, once replaced, the replacement will not last as long as the original unit. Instead of replacement, a valuable resource for such restoration on any older structure is the **National Park Service, Technical Preservation Services**:

https://www.nps.gov/tps/sustainability/energy-efficiency/weatherization/windows-doors.htm

08 00 00 - Openings

- **Window and Door Manufacturer's Association**

  2025 M Street, NW, Suite 800
  Washington, D.C. 20036-3309
  Phone: 202.367.1157

  http://www.wdma.com/

  The Window & Door Manufacturers Association defines the standards of excellence in the residential and commercial window, door and skylight industry and advances these standards among industry members

- **American Architectural Manufacturers Association** (AAMA)

  1827 Walden Office Square, Ste. 550
  Schaumburg, Illinois 60173-4268
  Phone: 847.303.5664
  Fax: 847.303.5774

  http://www.aamanet.org/

  "Since 1936, AAMA has stood as a strong advocate for manufacturers and professionals in the fenestration industry."

08 11 00 - Metal Doors and Frames

All products and materials need some level of maintenance, but in the rust belt locations, **thermally-broken aluminium frames** for glazing and door units are most common for non-residential construction *exterior* applications as compared to the **Hollow Metal (HM)** frames and door units. This is since HM is susceptible to rust, especially if the unit does not have a roof or canopy above. Yet, on the interior of a non-residential structure, the durable HM frames are very common for both doors and glazing. These HM door frames can be used for hinged wood or HM painted doors. The HM frames and the doors may need to have a fire rating while the glazed opening in the door or the "**Borrowed Lite**" will be determined by the building occupancy requirements listed in the applicable building code.

Hollow Metal door frames are not the most desirable aesthetically. Commonly, the only option is to paint them. However, one Williamsport, PA manufacturer has produced a magnetic wood door frame that easily snaps tightly into place - creating an easy solution with no mess. See below:

- **MagnaFit** - Magnetic Door Casings
  James Wood Company
  140 Catawissa Avenue
  Williamsport, PA 17701
  Phone: 570-326-3662

  http://www.jameswoodco.com/magnafit/

For a selection of **08 11 00** Metal Doors & Frames related LEED products, see:
http://www.arcat.com/divs/sec/sec081000.shtml

For a selection of **08 11 19** Stainless Steel Doors & Frames related LEED products, see:
http://www.arcat.com/divs/sec/sec081119.shtml

08 14 00 - Wood Doors

As you know, **wood door frames** are common in 1 and 2 family homes, places of worship and some non-residential applications, but be mindful of building code requirements for **Underwriters Laboratories (UL)** potentially required fire ratings. **http://www.ul.com/**

Along with this, it is important to properly specify required door swing direction for egress purposes along with door hardware, such as panic hardware and required door closures, at the openings of a fire rated wall.

To accommodate such code requirements, there are a few companies however that manufacture fire rated wood door frames, such as:

- **Navy Island, Inc.**
  275 Marie Ave. East,
  West St. Paul, MN 55118
  Phone: 651.451.4454
  Fax: 651.451.4484

  http://www.navyisland.com/products/wood-frames.html

Along with Fire Rated Wood Door Frames and Lites, **Navy Island** offers a variety of crafted products, including: Wood Acoustic Ceiling panels, planks and baffle systems; Acoustic Wall Panels; Architectural Wood Veneers; and Flush Wood Interior Doors.

- **Mohawk Doors**
  Masonite Architectural
  980 Point Township Drive
  Northumberland, PA 17857
  Phone: 570.473.3557

  http://www.mohawkdoors.com/

  *"**Mohawk** products are available with various environmental attributes to help your project achieve sustainable goals. Mohawk was one of the first door companies on the scene to support environmental doors. In fact, our doors can be found in the original U.S. Green Building Council headquarters located in Washington D.C."*

For wood **Pocket, Wall Mount, Bypass/Sliding, Multi-pass, Multi-fold Doors**, see:

- **L.E. Johnson Products, Inc.**
  2100 Sterling Avenue
  Elkhart, IN 46516
  Toll Free: 800.837.5664
  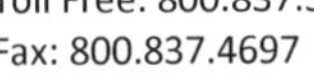
  Fax: 800.837.4697

  http://www.jhusa.net/

  L.E. Johnson Products, Inc. manufactures wood Pocket, Wall Mount, Bypass/Sliding, Multi-pass and Multi-fold Doors, parts and accessories for residential and light commercial applications. See the above for product information, videos, dealer locations or products can be purchased directly.

For a additional **08 14 00** Wood Door  related LEED products, see:
http://www.arcat.com/divs/sec/sec081400.shtml

08 16 13 - Fiberglass Doors

For doors and frames located within corrosive environments (such as an indoor pool that uses Chlorine at a School, Hotel or other similar applications), it is common to use **Fiberglass Reinforced Plastic (FRP)** and/or **stainless steel** (08 13 19) to prevent corrosion/rust. Such is also true within some manufacturing, pharmaceutical and medical environments for "cleanroom" requirements.

For a selection of **08 16 13** Fiberglass Door related LEED products, see:
http://www.arcat.com/divs/sec/sec081613.shtml

08 31 00 - Access Doors and Panels

Access Doors and Panels apply to products that typically provide access to plumbing, mechanical or electrical systems, via a door or panel commonly located in a ceiling or wall. This is an important feature item to plan for, specify and incorporate, and such doors or panels may need to be fire rated per the International Building Code (IBC), *Section* ***712 - Horizontal Assemblies,*** or per the International Residential Code (IRC), *Section* ***R807 - Attic Access*** requirements.

Use this website for *PremiumACCESS or* free *PublicACCESS* viewing of ***I-Codes*** *and* ***ICC Standards***:
https://codes.iccsafe.org/

For a selection of **08 31 00** related LEED products, see:
http://www.arcat.com/divs/sec/sec083100.shtml

08 32 00 - Sliding Glass Doors

Sliding Glass Doors typically offer daylighting and access to a patio, deck or balcony for Residences, Hotels, and other potential applications.

Sliding Glass Doors, French Doors, Folding Glass Doors and any other glass panel door allow the exterior and interior to blend - extending out to a well connected "outdoor room" - appropriate for many building types.

For a selection of **08 32 00** related LEED products, see:
http://www.arcat.com/divs/sec/sec083200.shtml

See other CSI Door Section headings for other various door options.

08 33 00 - Coiling Doors and Grilles

Coiling Doors, sometimes known as **roll-up doors**, offer overhead height flexibility in warehouses and other similar building occupancies as compared to a track supported overhead door. Coiling Doors and **Fire Shutters** can also be manufactured to provide the necessary **fire rated** separation between spaces and/or along fire rated corridors.

See: *McKeon Door - Auto-Set FSFD, Safescape T2000 and H200 Horizontal Fire Shutter*
https://www.youtube.com/watch?v=VQ-R5APt5qg 3:08

Coiling Grilles offer the same overhead height flexibility as Coiling Doors while providing sight and ventilation where required or desired (such as the the Coiling Grilles provided at the entry of retail tenant spaces within shopping malls), as compared to solid door panels.

For a selection of **08 33 00** related LEED products, see:
http://www.arcat.com/divs/sec/sec083300.shtml

08 37 00 - Garage Doors

Garage Doors, commonly designed as **overhead sectional doors**, are manufactured in panel sections with wheels mounted to them. To open the door, these wheels follow a track mounted to the inside face of wall, to an overhead position. These doors are the accepted norm in residential and other applications (including fire stations and some restaurants) where large exterior access is desired, and where forklift and/or storage rack height restrictions are <u>not</u> important. If there are such height restrictions, then a *Coiling Door* is best. *(See Section 08 33 00 Coiling Doors and Grilles)*

For a selection of **08 37 00** related LEED products, see:
http://www.arcat.com/divs/sec/sec083700.shtml

08 39 00 - Pressure-Resistant Doors

This CSI Section is divided into specific sub-sections that address **Watertight Doors; Hurricane-Resistant Doors; Tornado-Resistant Doors;** and **Blast-Resistant Doors.**

For a selection of **08 39 19** Watertight Door related LEED products, see:
http://www.arcat.com/divs/sec/sec083919.shtml

For a selection of **08 39 20** Hurricane-Resistant Door related LEED products, see:
http://www.arcat.com/divs/sec/sec083920.shtml

For a selection of **08 39 24** Tornado-Resistant Door related LEED products, see:
http://www.arcat.com/divs/sec/sec083924.shtml

For a selection of **08 39 53** Blast-Resistant Door related LEED products, see:
http://www.arcat.com/divs/sec/sec083953.shtml

08 41 00 - Entrances and Storefronts

It is important to understand the difference between a **Storefront** entrance and **Curtain Wall** system. A storefront system is one that fills an opening in a wall using extruded aluminium tubing that ultimately supports the glazing. Even though it is known as "Storefront", it is not limited to the front of stores. The wall opening(s) can be large or small, but will require the buildings' wall system to surround it for attachment. To do so, the storefront glass mullions (aluminium, steel or wood) are cut to length and fastened to the surrounding wall system, creating openings for the glass to be installed.

A Curtain Wall system is typically a full building height wall glazing system, with an extruded aluminium frame that is attached and hung, similar to a curtain. The buildings' structural system will then support the curtain wall from above and behind as it hangs on the outer face of the building. Where the curtain wall hangs in front of floor or mechanical systems, that do not want to be seen from the exterior. At these locations, **Spandrel Glass** is commonly used and is typically opaque, so to hide the items behind it. See this website (and 4:23 video) for related Spandrel Glass information:
http://educationcenter.ppg.com/glasstopics/spandrel_glass.aspx

Then, with the above, the importance of installation details. See, *Wall Cladding: Curtain Wall Systems.*
http://www.copper.org/applications/architecture/arch_dhb/arch-details/wall_cladding/curtainwall.html

- ## Rehabilitating Historic Storefronts

National Park Service
US Department of the Interior
Technical Preservation Services
Preservation Brief 11

https://www.nps.gov/tps/how-to-preserve/briefs.htm

- ## The Preservation of Historic Pigmented Structural Glass

National Park Service
US Department of the Interior
Technical Preservation Services
Preservation Brief 12

https://www.nps.gov/tps/how-to-preserve/briefs.htm

- ## Kawneer Company, Inc.

500 E 12th Street
Bloomsburg, PA 17815
Phone: 570.784.8000

http://www.kawneer.com/

*"(**Kawneer** is) continuing to lead the next generation sustainable buildings by focusing on innovative market solutions like high-thermal performance, hurricane resistance, and blast mitigation. From the unmatched **Kawneer doors**, the landmark **Trifab™ family of storefront framing systems**, the versatile **1600 Wall System™ curtain walls,** and the U.S. Department of Energy-backed **OptiQ™ Series windows**, our comprehensive product portfolio solves a myriad of high-, mid-, or low-rise application challenges."*

Kawneer has a Cradle to Cradle (C_2C) Product Registry.

- ## Pacific Architectural Millwork

1435 W. Pioneer St.
Brea, CA 92821
Phone: 562.905.3200
Fax: 562.694.6794

http://www.pacmillwork.com/cms/Portals/0/pdf-files/curtain-wall-system/PAM%20Curtain%20Wall.pdf

"As an alternative to standard aluminum storefront systems, Pacific presents the curtain wall system with wood interior. This system is ideal for applications where one desires a structural window wall with the warmth and richness of wood on the interior, such as high-end residential, boutique retail, or hospitality projects, with the durability and low-maintenance of aluminum or stainless steel on the exterior."

- ## Tubelite, Inc.

Corporate Headquarters
3056 Walker Ridge DR. NW, Ste. G
Walker, Michigan 49544
Toll Free: 800.866.2227

http://www.tubeliteinc.com/

Tubelite, Inc. is a leader in *"eco-efficient storefront, curtain wall and entrance systems"*, offering recycled aluminium for extruded products, *"with an industry leading content of post-consumer and pre-consumer material"*.

For additional **08 41 00** related LEED products, see:
http://www.arcat.com/divs/sec/sec084100.shtml

See also the following Division 08 Sections for related information:

08 43 33 - Folding Glass Storefronts

- ### Coverglass USA

  Showroom:
  2426 Newport Blvd, Ste E
  Costa Mesa, CA 92627

  Showroom:
  1287 Morena Blvd
  San Diego, CA 92110

  Toll Free: 800.317.2984

  http://www.coverglassusa.com/

Coverglass USA serves Southern California and is "*approved by the National Certified Testing Laboratories and are Title 24 compliant. We only use the latest generation of Low-E glass that has been tempered and laminated, with polished edges. Each system is custom built-to-order at our shop in San Diego, California.*"

- ### NanaWall Systems, Inc.

  100 Meadowcreek Dr, Ste 250
  Corte Madera, CA 94925
  Toll Free: 800.873.5673
  Phone: 415.383.3148

  http://www.nanawall.com/

NanaWall Systems, Inc. designs and manufacturers various Folding Glass Walls; Sliding Glass Walls; Frameless Glass Walls; and Screening Accessories for Residential and Commercial projects.

- ### Solar Innovations, Inc.

  31 Roberts Road
  Pine Grove, PA 17963
  Toll Free: 800.618.0669
  Phone: 570.915.1500
  Fax: 570.915.6083 or 800.618.0743

http://www.solarinnovations.com/folding-glass-walls/

"***Solar Innovations, Inc.*** *is a single source provider of residential and commercial folding, tilting, sliding, and stacking glass doors, walls, windows, and screens; skylights; greenhouses; conservatories; curtain walls; and more. If you don't see the* *product you're looking for, please contact your Solar Innovations, Inc. sales representative at: 800-915-1500."*

08 44 00 - Curtain Wall and Glazed Systems

Along with the manufacturers listed in *CSI Section **08 41 00 - Entrances & Storefronts***, see the below information:

- **Major Industries, Inc.**
  P. O. Box 306
  Wausau, WI 54402-0306
  Toll Free: 888.759.2678
  Phone: 715.842.4616
  Fax: 715.848.3336

  http://majorskylights.com/

  Major Industries, Inc. offers Translucent Curtain Wall Systems; Mixed Glazed Wall Systems; Canopies and Awnings; Translucent Skylights; Glass and Mixed Glass Skylights

For additional **08 44 00** related LEED products, see:
http://www.arcat.com/divs/sec/sec084400.shtml

08 45 00 - Translucent Wall and Roof Assemblies

- **EXTECH - Exterior Technologies, Inc.**
  200 Bridge Street
  Pittsburgh, PA 15223
  Toll Free: 800.500.8083
  Phone: 412.781.0991
  Fax: 412.781.9303
  http://www.extechinc.com/

  *"**EXTECH** specializes in the use of cellular polycarbonate for its host of benefits including daylighting, energy savings, superior impact resistance and insulation properties."*

- **Kalwall**
  1111 Candia Road
  P.O. Box 237
  Manchester, NH 03105
  Toll Free: 800.258.9777
  Fax. 603.627.7905

  http://www.kalwall.com/

  *"(**Kalwall** offers) the world's most advanced daylighting systems, (and it) is highly insulating and offers unrivaled Solar Heat Gain Control (SHGC)".* This company has been providing daylight solutions before daylighting was 'green'.

- **Panelite**
  8599 Venice Boulevard
  Los Angeles, CA 90034
  Phone: 323.297.0115

  http://www.panelite.us/

  Panelite manufacturers various products for daylighting and views designed for interior space partitions and exterior Insulating Glass Units (IGU) that provide benefits beyond typical glazing.

08 45 23 - Fiberglass Sandwich Panel Assemblies

For a selection of **08 45 23** related LEED products, see:

http://www.arcat.com/divs/divs_08.shtml

08 50 00 - Windows

As we all know, windows (sometimes affectionately known as '*Winders*'), are a very important element in design. Not only do they provide a quality design feature if proportioned properly to a buildings' exterior elevation, but they should also be sized, per code requirements so to provide the proper amount of daylight and ventilation, along with a secondary means of egress from bedrooms. See the beginning of Division 08 for related code information.

As it relates to the building code, if an existing residence has a bedroom window that is being replaced, yet the window size does not meet current building code egress requirements, typically as long as the replacement window is the same size, the window opening will not need to be altered. However, it is wise to check with the local building inspector.

See the below Sections for: Aluminum Windows; Steel Windows; Wood Windows; and Vinyl Windows.

08 51 00 - Aluminum Windows

- **Jeld-Wen, Inc.**

440 S. Church Street, Ste. 400
Charlotte NC 28202
Toll Free: 800.535.3936

http://www.jeld-wen.com/catalog/windows/premium-atlantic/aluminum

Jeld-Wen, Inc. manufactures Awning, Bay, Bow, Casement, Double-Hun, Fixed, Garden, and Single Hung style windows using various materials and finishes.

- 

J. Sussman, Inc.

109-10 180th Street
Jamaica, NY 11433
Phone: 718.297.0228
Fax: 718.297.3090

http://www.jsussmaninc.com/

"Family owned and run since 1906, ***J. SUSSMAN, INC.*** *is the oldest and largest manufacturer of custom windows and special aluminum extrusions for the stained and leaded art glass trade. We have the industry's largest selection of specialized church window systems (for stained and/or protection glass). In 2010* ***J. SUSSMAN, INC.*** *installed a solar photovoltaic array to offset 100% of its electrical energy load."*

- **Milgard Windows and Doors**

1010 54th Ave East
Tacoma, WA 98424
Toll Free: 800.645.4273

http://www.milgard.com/windows/aluminum/milgard-aluminum-awning-window

*"**Milgard** Aluminum windows have the flexibility to suit a multitude of projects ranging from custom homes to light commercial applications."*

- **Win-Door, Inc.**

7500 Amsterdam Drive
Orlando, FL 32832
Phone: 407.481.8400

http://beta.windoorinc.com/Home.aspx

Win-Door, Inc. manufactures alumimum high-performance, thermally broken, Certified impact resistant products that are available with Fin, Flange and Flush Framing options for new or renovation projects.

For additional **08 51 00** related LEED products, see:
http://www.arcat.com/divs/sec/sec085113.shtml

08 51 23 - Steel Windows

- **The Repair and Thermal Upgrading of Historic Steel Windows**

National Park Service
US Department of the Interior
Technical Preservation Services
Preservation Brief 13

https://www.nps.gov/tps/how-to-preserve/briefs.htm

- ## Hope's Windows, Inc.

84 Hopkins Ave, P.O. Box 580
Jamestown, NY 14702-0580
Phone: 716.665.5124
Fax: 716.665.3365

http://www.hopeswindows.com/hopes.shtml

*"**Hope's** custom-made steel windows and doors offer impeccable quality, unrivaled artistry and superior strength that complement any architectural theme. Unlike windows made from aluminum or wood, the depth and visual mass of a steel window are extremely narrow. Large glass lites and narrow sightlines result in better energy efficiency and unmatched aesthetics.*

Strength and durability of hot-rolled steel consistently outperform wood, aluminum and vinyl products. The life cycle and unique characteristics of Hope's custom products provide unparalleled value . . . (using) 100% recycled steel."

- ## Steel Window Institute (SWI)

1300 Sumner Avenue
Cleveland, OH 44115-2851
Phone: 216.241.7333
Fax: 216.241.0105

swi@steelwindows.com

*"We are an association of the leading manufacturers of windows made from either solid or formed sections of steel, and such related products as casings, trim, mechanical operators, screens, and moldings when manufactured and sold by members of the industry for use in conjunction with windows. **SWI** provides the public with general and technical information concerning the industry's products."*

For additional **08 51 23** related LEED products, see:
http://www.arcat.com/divs/sec/sec085100.shtml

08 52 00 - Wood Windows

Even though vinyl or aluminum clad wood windows offer the lure of being maintenance free, it is unfortunate that there has been class action lawsuits for such products that fail due to how moisture is then trapped behind the cladding, causing wood rot. It is better to restore the **existing wood windows** instead of replacing them, and if single-glazed, add **storm windows**. Studies have shown that a storm window offers similar energy savings as new windows with insulated glass. If the project involves new construction, it is better to purchase a non-clad, **solid wood window**; an all vinyl; all fiberglass; all aluminum; or an all steel window.

See this disturbing video from a home inspector that shows the results of **aluminum clad** and **vinyl clad** windows:
https://www.youtube.com/watch?v=YP0o9do4eSc
10min.

For related information, see also, **Openings: *Some additional information*** at the end of Division 08.

- **The Repair of Historic Wood Windows**
  National Park Service
  US Department of the Interior
  Technical Preservation Services
  Preservation Brief 9

  https://www.nps.gov/tps/how-to-preserve/briefs.htm

- **Duffy Hoffman, Preservation / SASH Consultant**
  Hoffman Preservation Restoration
  P.O. Box 922
  Elizabethville, PA 17023
  Phone: 215-262-0729

http://www.sashmasterduffy.com/default.aspx

"Please feel free in contacting us with questions about our DVD's, Books, classes and other services. Other services would include Restoration Consultant, Window SASH restoration, Maintenance and or Interior / Exterior Restoration."

To list a consultant is not typical, however, wood windows are unique and are threatened by what can be a quick decision to replace them. As mentioned earlier, studies have shown that storm windows are a better choice instead of the problems that have occurred with the installation of replacement windows. If it is not possible for Mr. Hoffman to be involved with your project, he may know of professionals closer to where your building is located. See CSI Section **08 80 00 - Glazing** for related information

- **Historic Wood Window Tip Sheet**
  National Trust for Historic Preservation
  1785 Massachusetts Ave., NW
  Washington, DC 20036
  Phone: 202-588-6000
  Fax: 202-588-6462

  http://www.dahp.wa.gov/sites/default/files/Wood-Windows-Tip-Sheet-July-2008.pdf

- **Preservation Pennsylvania**
  257 North Street
  Harrisburg, PA 17101
  Phone: 717-234-2310

www.preservationpa.org

Along with other resources, **Preservation PA** has written and offers their publication known as: *Considering the REPAIR, RETROFIT and REPLACEMENT of Historic Windows.*

A few examples of **non-clad, solid wood** windows:

- **Auralast - by Jeld-Wen, Inc.**
  440 S. Church Street, Ste. 400
  Charlotte NC 28202
  Toll Free: 800.535.3936

  http://www.jeld-wen.com/planning-projects/more-solutions/auralast-wood

  *"**JELD-WEN** wood windows, patio doors and door frames with **AuraLast Wood** can protect you from the expense, damage and inconvenience of decaying wood."*

- **JT Windows**
  9261 Independence Avenue
  Chatsworth, CA 91311
  Phone: 818.709.7950
  Fax: 818.709.4552

  http://www.jtwindows.com/

  *"We take great pride in our products and continue to work toward a common goal of providing the most exciting line of **premium solid wood** windows and doors in the industry."*

- **Menck Windows**
  77 Champion Drive,
  Chicopee, MA, 01020
  Phone: 866.296.3635

  http://www.menckwindows.com/products/solid-wood/

  *"**Menck Windows'** windows and doors are known for their energy efficiency, elegance, design range, comfort and quality. Menck USA, Inc. is committed to the best in worker safety and in environmentally sensitive and sustainable business practices."*

For additional **08 52 00** related LEED products, see:
http://www.arcat.com/divs/sec/sec085200.shtml

*See CSI Section **08 80 00 - Glazing** for related information*

08 53 13 - Vinyl Windows

Vinyl windows have come on strong over the past several years. As a replacement option be careful. If you own a home that is from the 1960's or older, the existing windows commonly can be restored. But, if they are replaced with vinyl, unfortunately such windows will not last more than 10-20 years. Unfortunately, this due to poor installation procedures, or the vinyl frame components break over time, or the seal breaks along the edge of the **glass spacer**. When the seal breaks, this causes the gas (commonly Argon or Krypton) to escape, which then allows air to enter between the glass panes, causing condensation to occur. See: **http://312windows.com/why-we-dont-install-vinyl-windows/**

If the vinyl windows are being installed in new construction, as with anything, do your homework!

- **Amsco Windows**
  1880 South 1045 West
  P.O. Box 25368
  Salt Lake City, UT 84125
  Toll Free: 800.748.4661

  http://www.amscowindows.com/

  It is understood that **AMSCO Windows** is the manufacturer used in the "***All-American Home***" described at the front of the book, built recently by **Anders Lewendal** and crew - in Bozeman, MT. AMSCO windows are available as vinyl or composite.

- **Wasco Windows**
  5070 N 124th Street
  Milwaukee, WI 53225
  Phone: 414.461.9900

  http://www.wascowindows.com/window-types/

  WASCO Windows manufacturers a variety of window styles, including **European style (Tilt & Turn)** or mullionless **(French)** windows. Along with these, they also offer Awning, Bay, Bow, Casement, Double-Hung, Fixed, and Slider windows.

For additional **08 53 13** related LEED products, see:
http://www.arcat.com/divs/sec/sec085313.shtml

08 60 00 - Roof Windows and Skylights

A **Roof Window** is a skylight that is able to open for ventilation. To open them, some are manually operated by a crank handle and some can be opened with the aid of a motor. Either way, be careful. Accidental water damage can happen if it is not closed properly prior to a storm! Also, be mindful about placing them too close to a plumbing vent stack, as there could be a horrible wafting down draft! Also, be careful with glass (or plastic) glazing on the roof. Depending on the geographical location, this could be the brunt of hail damage. A few other things to consider are glare from direct sun (at specific times of day), if transparent glass is used. Also, having penetrations in the ceiling/roof cause heat loss (or heat gain), since glazing typically will not provide the same about of insulation as the rest of the roof. *See CSI Section **08 80 00 - Glazing** for additional information.*

The above is also true with **Skylights**. They use fixed (non-operable) glazing, and can be installed as pre-packaged units, or they can be designed and manufactured to different shapes and sizes.

Tubular skylights offer a nice alternative. They provide the desired **daylighting** affect without the glare. And, these typically offer less heat loss and they can be easily adapted to an existing single-story space, with minimal effort and cost.

Each of the above can be appropriate for residential and non-residential application and are commonly used if this is the only way to provide daylight to an otherwise dark inner space or room. And, time and time again, studies have shown that daylight offers tremendous benefits. See link for related information:

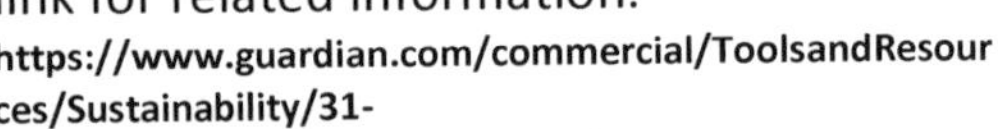

https://www.guardian.com/commercial/ToolsandResources/Sustainability/31-AdvantagesofEffectiveDaylighting/index.htm

At above website, select ***"White Paper - Benefits of Glass"***

A similar article, known as: ***The benefits of Natural Light,*** can be found in the *Architectural Lighting* magazine.

http://www.archlighting.com/technology/the-benefits-of-natural-light_o

08 61 00 - Roof Windows

For a selection of **08 61 00** related LEED products, see:

http://www.arcat.com/divs/sec/sec086100.shtml

08 62 00 - Unit Skylights

- **American Skylights, Inc.**

1218 Corporate Dr. E.
Arlington, TX 76006
Toll Free: 855.772.7401
Fax: 855.445.7282

http://www.americanskylights.com/

"***American Skylights** manufactures both unit skylights and roof systems for residential homes and for commercial applications. Our unit skylights come in a number of glazing options, including insulated glass, domes, and pyramids. Roof systems include barrel vaults, ridge mounts and ridge lites. Our skylights include standard and custom application.*"

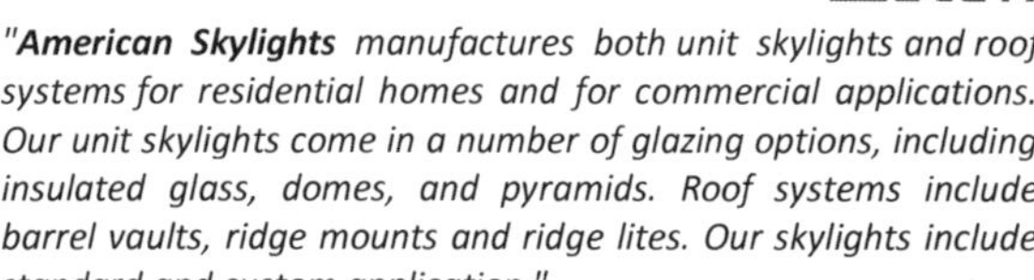

- **Sunoptics**

6201 27th St.
Sacramento, CA 95822
Toll Free: 800.289.4700
Fax: 916.395.9204

http://sunoptics.acuitybrands.com/about-us

"*Founded in 1978, **Sunoptics Prismatic Skylights** is a worldwide leader in high-performance daylighting solutions. Our mission is to harness the power of the sun to maximize the cost-effective energy savings of daylighting.*"

- **Wasco Lumira® Aerogel Skylights**

Eastern Manufacturing Facility:
85 Spencer Drive, Unit A, PO Box 559
Wells, ME 04090

Western Manufacturing Facility:
6645 Echo Avenue
Reno, NV 89506

Phone: 800-388-0293

http://www.wascoskylights.com/product/lumira-aerogel/

***Lumira® Aerogel** provides superior insulating capabilities. Made of a dry silica particulate, **Lumira® Aerogel** is completely recyclable, eco-friendly, non-combustible, and Cradle-to-Cradle certified. It is a lightweight insulation and daylighting solution that repels water, retains its properties under compression, and can enhance design options and aesthetics providing beautifully diffused full spectrum daylight inside a building.*

These skylights are UV stable and are made with "*multiwall polycarbonate glazing filled with **Lumira® Aerogel***".

For a list of **Lumira® Aerogel Partners,** see:
http://lumiradaylighting.com/contact_us.php

For additional **08 62 00** related LEED products, see:
http://www.arcat.com/divs/sec/sec086200.shtml

08 62 23 - Tubular Skylights

- **Elite Solar Systems Tubular Skylights**

310 E. Comstock Drive
Chandler, AZ 85225

Phone: 866-772-5418
Fax: 480-635-9767

http://elitesolarsystems.com/

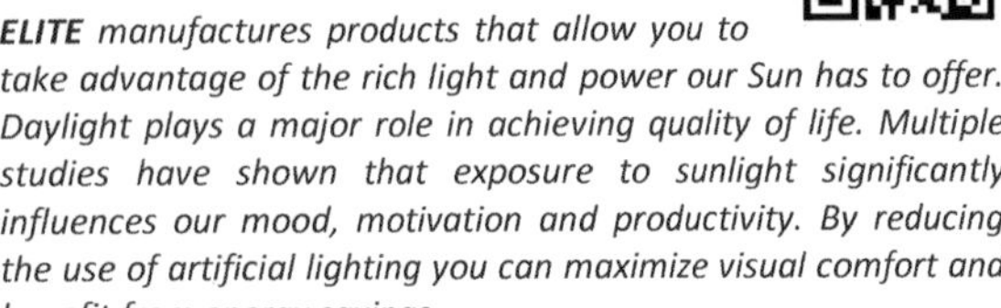

ELITE *manufactures products that allow you to take advantage of the rich light and power our Sun has to offer. Daylight plays a major role in achieving quality of life. Multiple studies have shown that exposure to sunlight significantly influences our mood, motivation and productivity. By reducing the use of artificial lighting you can maximize visual comfort and benefit from energy savings.*

- **Solatube International, Inc.**
  2210 Oak Ridge Way
  Vista, CA 92081-8341
  Toll Free: 888.765.2882

http://www.solatube.com/

"As a pioneer of the daylighting industry . . . ***Solatube Daylighting Systems****, are modular and easy to connect to ceiling systems. And unlike traditional skylights, they are designed to control the problematic aspects of sunlight. They reduce glare and inconsistent light patterns (that utilize patented optical technologies). They also screen infrared rays that can overheat interiors as well as ultraviolet rays that can fade furniture and fabrics."*

For a selection of **08 62 23** related LEED products, see:
http://www.arcat.com/divs/sec/sec086223.shtml

08 80 00 - Glazing

You have to admit that glazing (glass) is a fascinating natural product that is made from sand, sodium oxide and limestone that are melted together. And, as you know, when floated into sheets and cooled, then used as glazing, it offers a unique opportunity to separate us from the elements while also allowing us to enjoy the daylight and/or possible views to the great outdoors.

Glazing includes, but is not limited to clear window glass, decorative, mirrored & tinted glass, glazing accessories used to assemble the exterior or interior glass units, **glazing films**, (typically a tint product used to help shade the glass or strengthen it), Fire resistant glass, Ballistic resistant glass, and others that offer a comfortable and secure environment.

Glass today can range in overall thickness, but the most common application is made up of two panes of glass. Then, depending on the size of the window opening and wind loads will depend on how thick each pane of glass will be. But, generally thinking, the 2 panes of glass + the air space = 1" overall thickness.

However, the use of 3 panes of glass is becoming more and more common as well. Not only does this provide a greater thickness, but more weight as well. But, either way, the panes of glass then will have a **spacer** to separate the panes in order to create an "air space". With this there will be the **edge seal** that holds the glass together, but also seals in what is commonly either Argon or Krypton gas. Unfortunately, it is the edge seal that can fail which then ultimately allows the gas to escape and then condensation to form between the panes. The spacer material is an ongoing debate to which is best. And of course, technology is always changing. See related *article* from **Lawrence Berkley National Labs**, titled: *"Window Spacers and Edge Seals in Insulating Glass Units: A State-of-the-Art Review and Future Perspectives".*
See:
https://buildings.lbl.gov/sites/default/files/6122e.pdf

As previously discussed in Division 07, the glass insulating value is typically measured as the **U-Factor** and not **R-Value**. Since an R-Value and U-Factor are reciprocals of each other, one can find the equivalent so to compare one to the other. In other words, 1/R = U. Or, 1/U = R. So, with the above U= .35, the R-Value equivalent would be R= 1/.35 or *only* R-2.86. Then, with the above U= .18, the R-Value equivalent would be R= 1/.18 or *somewhat better* at R-5.56, yet still not good compared to the possible R-19 code minimum for wall insulation. We need to remember, that it is commonly understood that the majority (potentially up to 80%) of the heat is lost through the roof, and the balance of this through the walls. Then to understand that even a smaller percentage typically represents the fenestration, unless of course there is a full wall of glass, like a curtain wall.

Along with the glazing U-Factor, the **Solar Heat Gain Coefficient (SHGC)** is important. The SHGC varies depending on the type of glazing, the number of glass panes, along with if there are any glass coatings that help restrict the amount of solar radiation that actually enters the building through the entire window assembly as heat gain. See more information at:
http://www.commercialwindows.org/shgc.php

Insulated glass has evolved over time and it continues to do so. Prior to the 1950's, single pane glass was common. Since, studies have shown that when **storm windows** are added, the air space achieved between the two windows offers a similar, if not better insulation as that of common insulating glass of today. See this article from the State of Washington Department of Archeology & Historic Preservation, known as: ***"The TRUTH about - WINDOWS and Storm Windows Too!"*** : **http://www.dahp.wa.gov/sites/default/files/TheTruthAboutWindows.pdf**

Along with the above, there is various other supportive data and comments including the below grant supported study by www.architects.org, titled: *"**A Comparative Study of Cumulative Energy Use of Historical verses Contemporary Windows**"*. See: **http://www.architects.org/sites/default/files/Grant%20Final%20Report%2012-3-2010.pdf**

When it comes to shading and protecting the fenestration and other building elements from excessive heat gain and/or sun damage, a pent roof, overhangs, trellis, awnings, and the natural shade from trees are valuable. This website (and others) help to communicate awning and **overhang proportions.**
See: **http://www.susdesign.com/overhang/**

*See CSI Section **04 23 00 - Glass Unit Masonry,** for related products and information .*

08 81 00 - Glass Glazing

- **Pilkington - NSG Group**
  811 Madison Avenue
  Toledo, OH 43604-5684
  Toll Free: 800.221.0444
  Fax: 419.247.4517

  http://www.pilkington.com/en/us

  Pilkington manufacturers and distributes thermal insulation Low-E Glass, Solar Control Low-E Glass, Self Cleaning Glass, Mirror Coated Glass, Anti-Reflective Glass, Vacuum Insulated Glass, Noise Control Glass, High-Performance Tinted Glass, Fire Resistant Glass; Decorative Glass, Structural Glazing System Glass, and others, all for a variety of applications. The website offers a wealth of information, including videos.

- **PPG Industries, Inc.**
  GLOBAL HEADQUARTERS
  One PPG Place
  Pittsburgh, PA 15272
  Phone: 1.412.434.3131

  http://educationcenter.ppg.com/

PPG, once known as **Pittsburgh Plate Glass**, is "*Your single source for information on designing, specifying and building with commercial glass*", including spandrel glass. See the website "Education Center" for specific information and related videos.

- **PRL Glass Systems, Inc.**
  13644 Nelson Ave.
  City of Industry, CA 91746
  Toll Free: 800-433-7044
  Fax: 626-968-9256

  http://prlglass.com/

  PRL Glass Systems *"has the fastest leadtimes in the industry"* for Insulated Glass Units, Laminated Glass, Curtain Wall, Storefronts, Glass Doors, Accordion Bi-Fold Glass doors, and other glass specialty items for various applications.

For additional **08 81 00** related LEED products, see:
http://www.arcat.com/divs/sec/sec088100.shtml

08 81 13 - Decorative Glass Glazing

Decorative Glass would include **Frosted Glass**, **Leaded Glass**, **Stained Glass**, and **Beveled Glass**. An **Art Glass Window** would be the name for when one arranges the pieces of Stained or Beveled Glass in geometric patterns, like those designed by F. L. Wright that helped to define the *Prairie Style*, as well as other windows created during, or influenced by *The Arts & Crafts Movement in America.*

With such interest, even the Andersen Window-Door company offers such as an option. **https://www.andersenwindows.com/options-and-accessories/art-glass/**

This link shows the 1910 F.L. Wright *Robie House*:
http://architecture.uchicago.edu/locations/robie_house/ with images and video. 2:03

- **The Preservation and Repair of Stained and Leaded Glass**
  National Park Service
  US Department of the Interior
  Technical Preservation Services
  Preservation Brief 33

  https://www.nps.gov/tps/how-to-preserve/briefs.htm

For a selection of **08 81 13** related LEED products, see:
http://www.arcat.com/divs/sec/sec088113.shtml

08 87 13 - Window Film (Glazing Film)

- ### 3M Center

  Building 0235-02-S-27
  St. Paul, MN 55144-1000

  Toll Free: 866.499.8857
  Phone: 651.737.0196
  Fax: 651.737.3446

  http://solutions.3m.com/wps/portal/3M/en_US/Window_Film/Solutions/?WT.mc_id=www.3m.com/windowfilm

  3M manufacturers *"decorative, safety and energy saving window film solutions for your building".*

  *- **3M™ Safety & Security Film Ultra S600 Demonstration***
  https://www.youtube.com/watch?v=pG6nQZ_SwFA 1:41

  *- **3M™ Daylight Redirecting Film***
  https://www.youtube.com/watch?v=kM9pEUT735Y 3:26

  3M compares the benefits of prismatic film over light shelves

- ### LLumar Window Film

  Eastman Chemical Company
  575 Maryville Centre Drive
  St. Louis, MO 63141

  Toll Free: 800.255.8627
  Phone: 314.674.1000
  Fax: 314.674.1950

  http://northamerica.llumar.com/

  *"**LLumar** has been enhancing iconic, well-known buildings around the world for decades —with proven results. From Buckingham Palace to the Vatican to Chrysler World Headquarters, LLumar consistently exceeds the expectation of the most discerning customers."*

- ### Smart Tint

  http://www.smarttint.com/

  No address available so it is not certain this is a USA company. Yet, Smart Tint provides a Smart dimmable film (like Smart Glass), while being able to be applied to a variety of existing glass applications. This site provides sales and installation information.

- **Tap Plastics: How To Apply Window Film**
  https://www.youtube.com/watch?v=B22QdNHtb_8&ebc=ANyPxKqg0ouyoT1ULluqmHBZl1Ci6nT_dfqDzQCjqIb4OkaOtW1MHFt6yJfl4s6RZUthzZXstZ8NYDNtbqPkVGPXNf47WOrGFQ 8:45

Window Film: *Some additional information*

- **Time Warp - Window Glass**
  https://www.youtube.com/watch?v=hgGP372sVYQ 6:11
  This video show how glass can perform with and without window film.

08 88 36 - Switchable (Smart) Glass

- **Research Frontiers, Inc.**
  240 Crossways Park Drive
  Woodbury, NY 11797
  Toll Free: 888-SPD-REFR
  Phone: 516.364.1902
  Fax: 516.364.3798

http://www.smartglass.com/products/

*"**Research Frontiers** is the developer of (Suspended Particle Device) SPD-SmartGlass, electronically tintable window technology. Electronically tintable glass developed by Research Frontiers changes the tint of any window, sunroof or skylight by electrically aligning tiny particles in a thin film within the glass or plastic."*

- **View, Inc.**
  Headquarters
  195 S. Milpitas Boulevard
  Milpitas, CA 95035
  Phone: 408.263.9200

  Manufacturing
  12380 Kirk Road
  Olive Branch, MS 38654

http://viewglass.com/product/

"Forward thinking, energy conscious builders and architects who want the best for their occupants, View Dymanic Glass offers intelligent windows that maximize natural light and unobstructed views, while reducing heat and glare."

Switchable (Smart) Glass: *Some additional information*

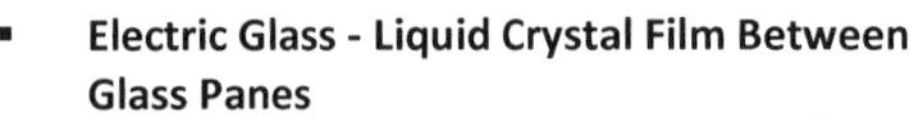

- **Electric Glass - Liquid Crystal Film Between Glass Panes**
  https://www.youtube.com/watch?v=l3UvwRHqEFA 1:26

- **Smart Glass - Privacy control and Display Applications**
  https://www.youtube.com/watch?v=Xrx5T-e1DA4&nohtml5=False 3:25

08 90 00 - Louvers and Vents

It should be understood that by design, there are ventilated and un-ventilated attic spaces. If the attic is ventilated, Louvers and Vents provide air movement. And with such, the ceiling plane (and the entire building envelope) should have a continuous air-barrier and insulation that prevents air movement from the conditioned space to the unconditioned attic space. With this, it is best to place all mechanicals below this thermal barrier to also eliminate air pressure differences. For a detailed discussion on this, See: **http://buildingscience.com/documents/digests/bsd-102-understanding-attic-ventilation**

- **Lamanco, Inc.**
  P.O. Box 519
  2101 West Main St.
  Jacksonville, AR 72076
  Toll Free: 800.643.5596
  Phone: 501.982.6511
  Fax: 501.982.1258

  Lomanco West, Inc.
  4450 Mohave Airport Road
  Kingman, Arizona 86401

  http://www.lomanco.com/

  Since 1946, **Lamanco, Inc.** has been manufacturing quality ventilation products, including the *"often imitated, but never duplicated"* Vari-Pitch Gable Vent. See website for other various roof and foundation vents and an on-line ventilation calculator.

For additional **08 90 00** related LEED products, see:
http://www.arcat.com/divs/sec/sec089000.shtml

Openings: *Some additional information*

- **Commercial Bldg. High-Performance Windows -** **http://www.commercialwindows.org/**
- **Large Roof Overhang: A Neglected Energy Saver -** **http://www.gohvacsales.com/static_docs/tech_tips/18_large_roof_overhang_a_neglected_energy_saver.pdf**
- **Energy Saving Storm Windows -** **http://energy.gov/energysaver/storm-windows**
- **Window Overhangs 101 -** **http://catalystarchitecture.com/SKETCH_PAD/Content/06/06/Green/Overhangs101.html**

- **Wood Exterior Door and Window improvements -** See Architectural Builders Supply - **http://absupply.net/** for weather stripping and other related products.

*See CSI Section **07 21 29 - Spray Foam Insulation,** for related products and information for Window Insulating Foam and Caulking/Sealing*

End of Division 08

Space for notes:

DIVISION 09 - Finishes

Finishes typically get a lot of attention. The space can be very well detailed and structurally dramatic, but it's those ***Finishes,*** that are the crowning touch - typically for floors, walls, and ceilings.

Unfortunately however, many finishes selected for various surfaces are often influenced by what is in style. And, if a product was installed years ago and still functions well, but because of its color or design, it may now look dated. As a result, it is no longer favorable, and is begging to be replaced. This unfortunately gets into the topic that was mentioned within this books' *Introduction* section, known as **Perceived Obsolescence**. This is where you and I need to be careful about which finishes and products we are specifying and/or installing. We really should select such products that are not influenced by trends and include products and colors that are timeless in any setting. Some examples to this could be the use of any natural material such as stone, brick, and wood. The point being, if a material is selected that is trendy, such a product will be seen as obsolete before it truly wears out. Thus, the product will easily be replaced and thrown away. A wall and trim that is painted is one thing. This can easily be repainted. But, what about the floor tile or carpet? And, what about wall paper?! The time it took to install such will now need to be duplicated to remove it. Then of course, dispose of such. This is where neutral / natural shades or colors are best. Such colors and materials are typically timeless and can color wise, be matched up with other items selected. By doing so, such items are not replaced before its time.

Another concept mentioned within the *Introduction* of this book is **Planned Obsolescence**. This is where we need to do our homework so to be confident knowing what products we specify and/or install, will be those that last. As we all know, the product might look wonderful immediately after installation, but . . ***Is the product safe? Will it last? / Is it durable?*** And, how does it compare with other similar products? One great resource that tests and compares various products is **Consumer Reports** - enabling us all to *"make the smart choice"*. See: **http://www.consumerreports.org/cro/index.htm**

As you may know, there are organizations and agencies like the **Consumer Products Safety Commission** that do their best to detect and protect against harmful products. **https://www.cpsc.gov/**

Yet unfortunately, products like ***Chinese Drywall*** or ***Chinese Laminate Flooring*** slip through - causing unforeseen problems and related costs. See: *CNN - Danger of Some Laminate Wood Flooring Was Underestimated, Report Says* **http://www.cnn.com/2016/03/22/health/health-effect-laminate-wood-flooring/**

Along with Consumer Reports and the Consumer Products Safety Commission, there is additional information throughout this book.

Again, those questions are:

1. Is the product safe to use and/or install? With the current topic of *Division 09 - Finishes*, it would be wise to consult the International Building Code (IBC), ***Chapter 8 - Interior Finishes,*** and the International Residential Code (IRC), ***Chapter 7 - Wall Covering.*** Ask, has it proven itself over time? Will it be a toxic nightmare? Natural materials / products are usually safe.

Use website to access the *I-Codes,*

PremiumACCESS or the Free PublicACCESS: **https://codes.iccsafe.org/**

2. Will the product last / is it durable? This is where research comes in. To seek out input from those that have installed such and along with the end users for durability and maintenance. Do not simply rely on what is found on the Internet.

In the end, if the product does not hold up to its expectations, or is no longer is trendy as previously discussed, unfortunately it will be replaced and tossed into the landfill. This is wrong. And, as you know, not sustainable!

*- See the topics of "Planned Obsolescence", "Perceived Obsolescence", and "What Makes A Product Green?", found within the **Introduction**, for related information.*

*- For information related to Drywall, see CSI Section **09 20 00 - Plaster and Gypsum Board***

*- For information related to Laminate Flooring, see CSI Section **09 62 19 - Laminate Flooring***

09 01 20.91 - Plaster Restoration

- **Repairing Historic Flat Plaster - Walls & Ceilings**
  US Department of the Interior
  Technical Preservation Services
  Preservation Brief 21

  https://www.nps.gov/tps/how-to-preserve/briefs.htm

- **Repairing Historic Ornamental Plaster**
  US Department of the Interior
  Technical Preservation Services
  Preservation Brief 23

  https://www.nps.gov/tps/how-to-preserve/briefs.htm

- **Plasters Manual**
  Portland Cement Association
  5420 Old Orchard Road
  Skokie, Illinois 60077-1083
  Phone: 847.966.6200
  Fax: 847.966.9781

  https://archive.org/details/PlasterersManual

  This beautiful electronic book describes the professional craft of plaster and stucco finishes while being supplemented with photographs and hand drawn details.

09 20 00 - Plaster and Gypsum Board

- **Gypsum Association**

  6525 Belcrest Road, Suite 480
  Hyattsville, MD 20782
  Phone: 301.277.8686
  Fax: 301.277.8747

  https://www.gypsum.org/about/

  *"Since 1930, the **Gypsum Association** has promoted the proper use and installation of gypsum products while advancing the development, growth, and general welfare of the gypsum industry in the United States and Canada."*

- **AirRenew**

  CertainTeed Corporation
  P.O. Box 860
  Valley Forge, PA 19482
  Professional: 800-233-8990
  Consumer: 800-782-8777

  http://www.certainteed.com/airrenew
  with 3:19 video

  *"The **AirRenew®** family of products are the only gypsum boards that actively clean the air. They feature the industry's first formaldehyde-absorbing technology which actively removes formaldehyde from the air, converting it into a safe, inert compound, improving indoor air quality for generations."*

- **American Clay Enterprises, LLC**

  2418 2nd Street SW
  Albuquerque, NM 87102
  Toll Free: 866.404.1634
  Phone: 505.243.5300

  http://www.americanclay.com/

  *(The **American Clay**) natural earth plaster products, manufactured in Albuquerque, New Mexico, are a healthy and creative solution for beautiful interiors. Our mission is to bring universal awareness to the value of environmentally conscious products to indoor spaces! Our gorgeous, sustainable clay plasters offer a non-toxic alternative for residential and commercial interior wall coverings and they are unparalleled in quality, health, beauty and value.*

  Applying American Clay - Introduction video:
  https://www.youtube.com/watch?v=QoM3Agh-RYE 2:52

- **Dens® Brand Fiberglass Mat Gypsum Panels**

  Georgia-Pacific Building Products
  133 Peachtree Street NE
  Atlanta, GA 30303
  Phone: 404.652.4000

  http://www.buildgp.com/dens-fiberglass-mat-gypsum with 2:33 video

  "Georgia-Pacific Gypsum introduced fiberglass mat gypsum panel technology to the construction industry 30 years ago.

"Glass mat" technology enabled important changes in how the building envelope can be fortified to resist damage from moisture and mold, during and after construction.

Dens® Brand Gypsum Panels can help AEC professionals achieve goals for sustainable design and construction, can help contractors manage scheduling pressures, and can help mitigate the risk of remediation due to moisture damage."

- **EcoRock** and **Quiet Rock**

  Serious Materials / Serious Energy, Inc.
  1250 Elko Drive
  Sunnyvale, CA
  Toll Free: 800.797.8159 (now PABCO, see spec)

  http://www.seriousenergy.com/

  What happened to Serious Materials / Serious Energy, and its various products, including Serious Windows and Doors?

  - **http://www.greentechmedia.com/articles/read/serious-energy-in-serious-trouble**

  And, what Happened to the ***Green New Deal***? *(Obama, 2008 economy, Solyndra, and how many others?!?)*

  - **http://www.truth-out.org/news/item/12610-what-happened-to-the-green-new-deal**
  - **https://www.washingtonpost.com/politics/specialreports/solyndra-scandal/**

  *- See CSI Section **09 30 15 - Glass Tiling,** for related information.*

  EcoRock: (2005, 85% recycled material + 20% of the energy)
  Link to: **Popular Science Article - Dec 2008**

  TED Talk: Kevin Surace, EcoRock (2005)
  https://www.ted.com/talks/kevin_surace_fixing_drywall_to_heal_the_planet?language=en#t-14329 3:10

  Link to: **QuiteRock 3-Part Specifications prior to PABCO**
  QuiteRock: (2008), since sold to PABCO Gypsum (see below)

- **Forever Board**

  Forever Board, Inc.
  1 Buffalo River Place
  Buffalo, NY 14210
  Phone: 716.213.8640 or 716.213.8637
  Fax: 716.235.8445

  http://www.foreverboard.net/ (with 6:44 video)

  *"The main features of (**FOREVER BOARD** include) Magnesium Oxide (which is) fire resistance, moisture resistance, mold and mildew resistance, and strength. Magnesium Oxide is a refractory material, which means it is physically and chemically stable at high temperatures and gives our boards freeze/thaw stability. This provides the fire resistance in our boards. There are no nutritious material in our boards, making them resistant to mold, fungus and insects."*

- ## QuiteRock

  PABCO Gypsum
  37851 Cherry Street
  Newark, CA 94560

  Toll Free: 800.797.8159 (Number previously for *Serious.* See above 3-part Spec. Link)
  Phone: 408.541.8000
  Fax: 408.715.2560

  http://www.quietrock.com/

  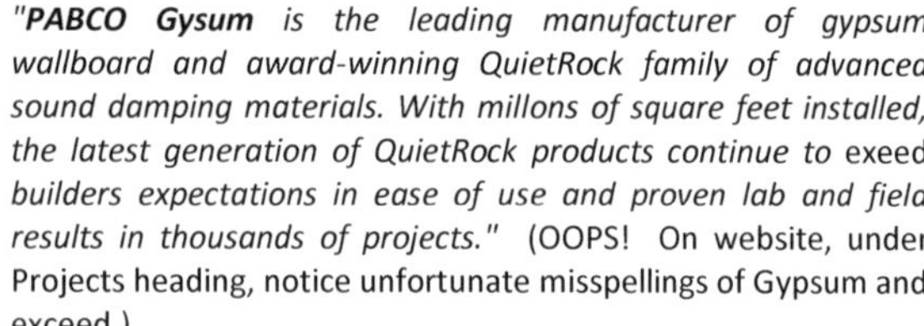

  *"**PABCO Gysum** is the leading manufacturer of gypsum wallboard and award-winning QuietRock family of advanced sound damping materials. With millons of square feet installed, the latest generation of QuietRock products continue to* exeed *builders expectations in ease of use and proven lab and field results in thousands of projects."* (OOPS! On website, under Projects heading, notice unfortunate misspellings of Gypsum and exceed.)

- ## Silent FX

  CertainTeed Corporation
  P.O. Box 860
  Valley Forge, PA 19482
  Professional: 800.233.8990
  Consumer: 800.782.8777

  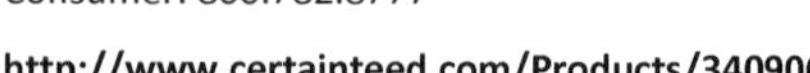

  http://www.certainteed.com/Products/340906

  SilentFX® *is a noise-reducing gypsum board specifically designed for systems requiring high STC ratings where acoustic management is needed. SilentFX gypsum board is enclosed in a 100% recycled moisture and mold resistant face and back paper.*

- ## TAKCOAT Platinum™ Lime Plaster

  Limeworks.us
  3145 State Road
  Telford, PA 18969
  Phone: 215.536.6706
  Fax: 215.453.1310

  http://www.limeworks.us/home.php

  *"**TAKCOAT Platinum™** is a specially formulated lime plaster base coat designed to be installed on sound interior/exterior applications without the need for first priming a wall."* Other benefits include: *"Single coat or base coat plaster; Environmentally friendly; VOC free; LEED; Post consumer recycled content; Breathable; No lath; No acrylic bond; and No sanding".*

  Limeworks.us - **How to apply Lime Plaster to Sheetrock:**
  https://www.youtube.com/watch?v=rpkw-a2Me9c 3:00

- ## ThermaRock XI Drywall

  Serious Materials / Serious Energy, Inc.
  1250 Elko Drive
  Sunnyvale, CA

  http://www.businesswire.com/news/home/20080916005592/en/Materials-Introduces-ThermaRock-TM-Energy-Saving-Drywall

  ThermaRock XI Energy Saving Drywall is another product by the failed Serious Materials company. Not sure if this product was ever manufactured, or if so, where it was manufactured. It is listed here though as inspiration. As of printing, it appears that this product was not sold to another company as QuiteRock was.

- **Trim-Tex Drywall Products**

  Trim-Tex Main
  3700 W. Pratt Ave
  Lincolnwood, IL 60712
  Phone: 800-874-2333
  Fax: 800-644-0216

  Trim-Tex West
  1487 N. Main St.
  Orange, CA 92867
  Phone: 800-874-2333
  Fax: 800-644-0216

  http://www.trim-tex.com/education-center/

  "You have got to see it to believe what you can do with ***Trim-Tex rigid vinyl beads.*** *Transform drywall scraps into beautiful Drywall Art. Upgrade any interior by adding wainscotting, window casings, ceiling medallions, and more made from drywall waste that would otherwise be hauled to a landfill. Reduce waste and increase profits on any job."*

- **United States Gypsum Corp.** (USG)

  60 PPL Road
  Washingtonville, PA 17821
  Toll Free: 800.874.4968

  https://www.usg.com/content/usgcom/en/products-solutions/products/wallboard.html

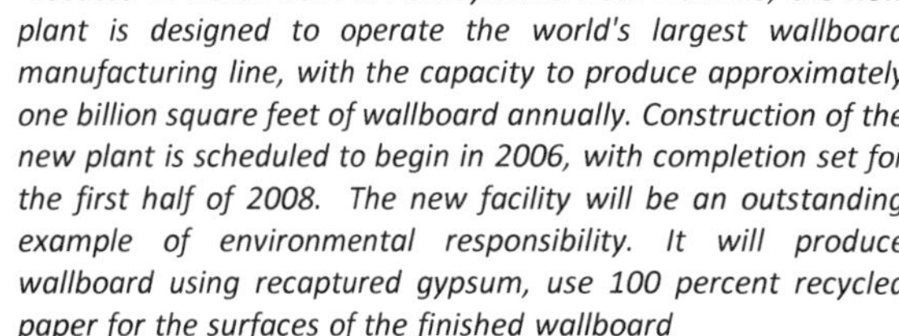

"Located in North Central Pennsylvania near Danville, the new plant is designed to operate the world's largest wallboard manufacturing line, with the capacity to produce approximately one billion square feet of wallboard annually. Construction of the new plant is scheduled to begin in 2006, with completion set for the first half of 2008. The new facility will be an outstanding example of environmental responsibility. It will produce wallboard using recaptured gypsum, use 100 percent recycled paper for the surfaces of the finished wallboard products, recycle 100 percent of its production waste and feature a closed-loop liquid effluent system, which translates to zero discharge into nearby waterways."

See:**http://www.gypsumtoday.com/news/viewnews.pl/id=577**

For additional **09 20 00** related LEED products, see:
http://www.arcat.com/divs/sec/sec092000.shtml

Gypsum Board: *Some additional information*

- As with the unfortunate Chinese Laminate Flooring problem, the **Chinese Drywall** problem became a class-action lawsuit as well.

  https://topclassactions.com/lawsuit-settlements/lawsuit-news/2142-80m-chinese-drywall-class-action-settlement-reached/

- To attach drywall / Gypsum Board to non-structural metal framing, see *CSI Section **09 22 00 - Supports for Plaster and Gypsum Board***

09 22 00 - Supports for Plaster and Gypsum Board

This section is for **non-structural metal framing** (Cold-Formed metal hat-channel furring, studs, and joists). For structural metal framing, see *CSI Section **05 40 00.***

- **ClarkDietrich Building Systems**

  Corporate Office
  9100 Centre Pointe Drive, Suite 210
  West Chester, OH 45069
  Phone: 513.870.1100
  Fax: 513.870.1300

  http://www.clarkdietrich.com/

  ClarkDietrich engineers and manufacturers **cold-formed structural steel framing** products and systems, including but not limited to **joists, studs, BIM Services,** and **LEED Services** for projects of all sizes.

- **Steeler Construction Company**

  Corporate Office
  10023 MLK Jr. Way South
  Seattle, WA 98178
  Phone: 800.275.2279
  Local: 206.725.2500
  Fax: 206.725.1300

  http://www.steeler.com/

  Steeler Construction Company *"manufactures and distributes high quality steel framing products for just about any project need"* - that can provide **LEED credits** as with the **ELITE Steel Framing System™**.

  Other Steeler locations include: Tacoma, WA; Spokane, WA; Portland, OR; Tigard, OR; Sacramento, CA; San Francisco, CA; Bakersfield, CA; San Diego, CA; Phoenix, AZ; Tucson, AZ; and Delta, BC.

For additional **09 22 00** related LEED products, see:
http://www.arcat.com/divs/sec/sec092200.shtml

09 23 00 - Gypsum Plastering

For a selection of **09 23 00** related LEED products, see:
http://www.arcat.com/divs/sec/sec092300.shtml
and, ***CSI Section 09 01 20.91 - Plaster Restoration***

09 24 00 - Cement Plastering

- **The Preservation and Repair of Historic Stucco**
  US Department of the Interior
  Technical Preservation Services
  Preservation Brief 22

  https://www.nps.gov/tps/how-to-preserve/briefs.htm

For additional information on **09 24 00** related LEED products, including *Adobe Finish, Cement Stucco, Cement Parging*, and *Lime Based Plastering*, see: *CSI Sections* ***09 01 20.91; 09 20 00; 09 23 00,*** *and Section* ***04 05 13*** - Along with: **http://www.arcat.com/divs/sec/sec092400.shtml**

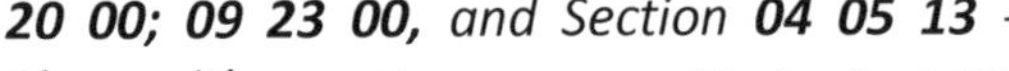

09 30 13 – Ceramic and Porcelain Tiling

This section can apply to both walls and floor tile finishes. Make sure to browse all the following sections, including *CSI Section* ***09 60 00.***

As we know, no floor or wall finish is maintenance free, however, ceramic tile, generally speaking, offers a durable finish that has been used for centuries in various residential and non-residential applications. Along with this, ceramic tile typically is made in various shapes, sizes and colors to offer limitless pattern variations. And, with all the grout colors options, the grout can accent or blend with the tiles selected.

Ceramic Tile for walls and floors is made from white or red clay that is then fired in a kiln, commonly with a glazed finish.

Porcelain Tile for walls and floors is made from porcelain clays that offer a denser, and more durable product compared to ceramic tile - while also offering a lower water absorption rate, which allows them to be more frost resistant if used for exterior applications.

Common tile manufacturers are:

- **Acme Brick Tile & Stone Company**
  3024 Acme Brick Plaza
  Fort Worth, TX 76109
  Toll Free: 800.792.1234
  Phone: 817.332.4101
  Fax: 817.390.2404

  http://www.acmebricktileandstone.com/

  *"**Acme Brick Tile & Stone** offers a variety of porcelain, ceramic and stone flooring. Decorative accents, hand painted wall tiles, glass and stone mosaics and metal accents are also available."*

- **American Olean Tile Co.**
  7834 C. F. Hawn Freeway
  P. O. Box 17130
  Dallas, TX 75217
  Toll Free: 888.AOT TILE
  Phone: 214.398.1411
  Fax: 215.822.1353

  http://americanolean.com/index.cfm

  American Olean®, (owned by Dal-Tile) offers a variety of porcelain, quarry, glass, metal, mosaic, medallions & stone tile.

- **Daltile® Corporation**
  7834 C.F. Hawn Freeway
  P. O. Box 170130
  Dallas, TX 75217

  Toll Free: 800.933.TILE
  Phone: 214.398.1411
  Fax: 214.309.4584

  http://www.daltile.com/

  Daltile®, (owned by Dal-Tile) offers a variety of porcelain, quarry, glass, metal, mosaic, accent & stone tile.

- **Preserving Historic Ceramic Tile Floors**
  US Department of the Interior
  Technical Preservation Services
  Preservation Brief 40

  https://www.nps.gov/tps/how-to-preserve/briefs.htm

For additional **09 30 13** related LEED products, see:
http://www.arcat.com/divs/sec/sec093013.shtml

09 30 15 – Glass Tiling

- **Fireclay Tile**
  901 Brannan Street
  San Francisco, CA 94019
  Phone: 408.275.1182

https://www.fireclaytile.com/

*"At **Fireclay,** our goal is to make amazing tile exactly the way you want it. We are unique from anyone else in that we make every tile to order using recycled materials and sustainable manufacturing practices, all right here in California."*

- ## Glass For Good

c/o Shares, Inc.
1611 South Miller Street
Shelbyville, Indiana 46176
Phone: 877.398.8218
Fax: 317.398.7838

http://glassforgood.com/

*"**Glass for Good** is a division of **Shares, Inc.**, which is a private not-for-profit agency that provides job and basic life skills training for adults with developmental disabilities. Through the **Glass for Good** / glass rock program, our associates learn the importance of recycling glass, while earning a wage for packaging the recycled glass material."*

For additional **09 30 15** related LEED products, see *CSI Section* ***09 30 13*** and:
http://www.arcat.com/divs/sec/sec093015.shtml

Glass Recycling: *Some additional information*

At their website, **https://www.fireclaytile.com/blog/full/At-The-Factory-Making-Glass/**, Fireclay Tile and others have benefited from the use of unused **Solyndra** solar photovoltaic (PV) glass tubes along with other creative uses such as creating modern art. See:
https://pjmedia.com/zombie/2012/8/20/where-our-500-million-went-solyndra-glass-tubes-used-as-modern-art/

*- See CSI Section **09 20 00 – Plaster and Gypsum Board,** for related products and information (related to the failed **Green New Deal effort**.)*

09 30 16 – Quarry Tiling

- ## American Olean Tile Co.

7834 C. F. Hawn Freeway
P. O. Box 17130
Dallas, TX 75217
Toll Free: 888.AOT TILE
Phone: 214.398.1411
Fax: 215.822.1353

http://americanolean.com/series.cfm?series=31&c=47

*Durable and long-wearing, **Quarry Naturals'** slip-resistant cushioned edge and unglazed surface make them eminently* *suitable for outdoor patio and pool areas as well as floors and walls.*

- **Daltile® Corporation**

7834 C.F. Hawn Freeway
P. O. Box 170130
Dallas, TX 75217
Toll Free: 800.933.TILE
Phone: 214.398.1411
Fax: 214.309.4584

http://products.daltile.com/catalog.cfm?c=12

Daltile® Quarry and Saltillo Tile - *"Appealing to both classic and contemporary tastes, Daltile's Quarry collections have a beautiful and stylish breadth of product to arouse the most inspired design."*

For additional **09 30 16** related LEED products, see:
http://www.arcat.com/divs/sec/sec093016.shtml

09 30 23 - Glass Mosaic Tiling

*For a selection of **09 30 23** related LEED products, see CSI Section **09 30 15** and:*
http://www.arcat.com/divs/sec/sec093023.shtml

09 30 29 - Metal Tiling

For a selection of **09 30 29** related LEED products, see:
http://www.arcat.com/divs/sec/sec093029.shtml

09 30 33 - Stone Tiling

For a selection of **09 30 33** related LEED products, see:
http://www.arcat.com/divs/sec/sec093033.shtml

09 30 50 - Tile Setting Materials & Accessories

- **LATICRETE International, Inc.**

1 LATICRETE Park
North Bethany, CT 06524
Phone: 203.393.0010, ext. 235

Toll Free: 800.243.4788, ext. 235
Fax: 203.393.1684

http://www.laticrete.com/

Latricrete Int'l manufacturers of a variety of surface preparation, shower systems, adhesives, grouts, sealers along with paver sand and paver sealers.

For additional **09 30 50** related LEED products, see:
http://www.arcat.com/divs/sec/sec093050.shtml

To follow industry / professional installation methods, it should be noted that floor, wall and ceiling tiles are typically centered within the space and/or surface, for design aesthetics and installation reasons.

09 50 00 – Ceilings

Ceilings, like walls and floors play an important part of a space aesthetically and acoustically. As we can all imagine, aesthetics, ties into the Perceived Obsolescence topic described at the beginning of Division 09, and elsewhere. Acoustics on the other hand never goes out of style, and really becomes a specialty profession. And, hiring an **Acoustic Engineer** is valuable for various projects and the needs and the end result will vary depending what the intended use will be for the space, such as a within a large auditorium, or a small sound recording studio, and potentially all the various types of venues in between.

Generally speaking, it is important to remember, that hard surfaces bounce sound and soft surfaces (including people in a room), absorb sound. Then, depending on what part of the building is receiving a certain ceiling finish, the **Fire Rating** of the material (Class A, B or C, for fire spread and smoke generation, per Code), along with the **Sound Transmission Class (STC)** ratings are important to consider.

A related topic to the fire and smoke would be **Indoor Air Quality (IAQ).** Keep in mind that if and when a fire were to break out, and, if there are occupants within the structure, it is the smoke that people are commonly treated for, so it would make sense to select products that are not harmful if burned or otherwise.

For an overview on code related fire and smoke requirements, see IBC, Chapter 8 - Interior Finishes, as previously mentioned, at the beginning of Divison 09 of this book.

http://publicecodes.cyberregs.com/icod/ibc/2009/icod_ibc_2009_8_section.htm

For an overview on STC, see:
http://www.soundproofingcompany.com/soundproofing101/understanding-stc/

With the aesthetic topic, ceilings, as we know can be the typical hard ceiling (plaster or gypsum board) or an acoustic (suspended / lay-in) tile system. But, remember that there are other options as well, including wood, metal, and plaster finishes, as well as various soffit systems, integrated light cove systems, luminous ceiling system, and/or exposing the mechanicals along with the roof deck or floor above - to name a few. Which ever ceiling system is selected, this is where fire separation per Code needs to be addressed, especially if another tenant or change of occupancy is above or below.

And, similar to what was mentioned in *CSI Section **09 30 50***, it should be noted that full room lay-in suspended ceiling grid systems are typically centered within the space, for design and installation reasons, as represented in a *Reflected Ceiling Plan* drawing. This drawing not only shows the anticipated ceiling finish / ceiling tile pattern, but also other possible ceiling features, including but not limited to soffits, lighting, HVAC diffusers and returns.

- **Armstrong World Industries**
  824 Linden Street
  Bethlehem, PA 18018
  Phone: 610.865.6700

  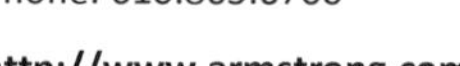

  http://www.armstrong.com/

  *"**Armstrong Ceilings** is proud of our heritage and the growing portfolio of ceiling and wall products and programs designed to support sustainable building. It's our goal in all areas of our business to continually reduce our environmental footprint."*

- **Pinta Acoustic, Inc.**
  2601 49th Avenue N #400
  Minneapolis, MN 55430
  Phone: 612.355.4200

  Toll Free: 800.662.0032 (US and Canada only)
  Fax: 612.355.4299

  http://www.pinta-acoustic.com/

*"**Pinta Acoustic** is part of the Pinta Elements Group—working together to serve the European and North American markets. Over the last 50 years, the company has gained extensive experience with acoustic and noise control products."*

For additional **09 50 00** related LEED products, see:
http://www.arcat.com/divs/sec/sec095000.shtml

09 60 00 – Floorings

It is important to remember that any surface, including the floor will be worth the effort if it the sub-surface is properly prepared. Such preparation, self-levelling and adhesive products, including related underlayment product(s), and any other preparation or procedures needed should be communicated in the flooring specifications, so to meet the manufacturer's flooring warranty. See *CSI Section 09 30 50* along with the following sections for related information.

09 60 13 - Acoustic Underlayment

- **Earth Weave Carpet Mills, Inc.**
  P.O. Box 6120
  Dalton, GA 30722
  Phone: 706.278.8200
  Fax: 706.278.8201

  http://www.earthweave.com/padding/

  *"Founded in 1996, **Earth Weave** was the first carpet manufacturer in the United States to introduce a running line product that incorporates only 100% natural materials,* (including **ENERTIA™**,) *100% natural wool carpet padding".*

- **Homasote Company**
  932 Lower Ferry Road
  West Trenton, NJ 08628
  Toll Free: 800.257.9491

  http://www.homasote.com/

  *"For 100 years Homasote®, has been the only manufacturer of its kind in North America. **Homasote's** environmentally-intelligent building and industrial packaging products made of 98% recycled materials are truly unique, both in their manufacturing processes and their dependable fiberboard construction performance."*

- **Whitfield Natural Textiles**
  Whit-Tex, Inc.
  611 Oxford Street
  Dalton, Georgia 30720
  Toll Free: 800.521.1950

Fax: 706.278.8440

https://sites.google.com/a/whit-tex.com/whit-tex/whisper-wool-technical-information

Whit-Tex, Inc. manufactures a natural acoustic carpet underlayment known as **Whisper Wool.** (It) "*is manufactured by means of preparing wool fiber and processing those fibers through a needle punch machine. All materials used shall be new, of good quality and without defects that would lessen the quality of the product.*"

See more info at:
http://www.greendepot.com/greendepot/product.asp?dept_id=&pf_id=WHISWOOL&

For additional **09 60 13** related LEED products, see:
http://www.arcat.com/divs/sec/sec096013.shtml

09 60 50 - Flooring Adhesives

Flooring Adhesives have improved drastically over the years, removing many Volitile Organic Compounds (VOCs). It was many adhesives and some finishes that were made with various chemicals that off-gassed and caused what is/was known as **Sick Building Syndrome.** However, some adhesives are not strong enough to do the job even though they have low VOCs or even Zero VOCs as witnessed in the movie *The Greening of Southie.* This was a documentary about The Macallen Building, a high-end residential structure that was built as LEED Gold, in South Boston, MA. The problem was that the adhesive that was originally specified and installed caused the bamboo flooring to buckle. The result was to remove and discard much of the flooring - causing a tremendous problem and expense. This is not acceptable, nor sustainable.

- **Dri-Tac Flooring Products, LLC**
  60 Webro Road
  Clifton, NJ 07012
  Toll Free: 800.394.9310
  Phone: 973.614.9000
  Fax: 973.614.9099

  http://www.dritac.com/home.php

  *"With over 59 years manufacturing in the United States and providing the domestic and international markets with innovative, proven eco-friendly adhesive technology, **DriTac Flooring Products LLC** is one of the most experienced manufacturers of adhesives in the market today."*

For additional **09 60 50** related LEED products, see: **http://www.arcat.com/divs/sec/sec096050.shtml**

09 62 19 - Laminate Flooring

Laminate Flooring is easily installed as a **'floating floor'**. The product typically snaps into place on top of a foam pad. However, based on observing several professional installations, the seams can easily raise if exposed to moisture, such as at an entrance door, a plumbing fixture, or being poorly maintained with wet mopping. There can be a hollow clicking sound as one walks across it, even after the foam pad is in place. Just like with any product, it is important to understand the materials used in manufacturing. Some laminate flooring made in China was the result of an unfortunate class action lawsuit. See: **http://www.classaction.org/lumber-liquidators**

With your review of the link, I would expect that you will agree that this is not sustainable.

EPA - Questions and Answers Regarding Laminate Flooring:
https://www.epa.gov/formaldehyde/questions-and-answers-regarding-laminate-flooring

09 62 50 - Indoor Athletic Flooring

- **Gerflor USA**
  595 Supreme Drive
  Bensenville, IL 60106
  Phone: 877.437.3567
  Fax: 847.394.3753

  http://www.gerflorusa.com/products/sports-products.html

  *"**Gerflor** products are inherently and endlessly recyclable. The Gerflor "Second Life" program provides full recycling services from removal to sorting to storage, to repurposing into other products such as road cones. For the London Olympic Games, we took back nearly 475,000 sqft of Gerflor flooring and recycled all 75 tons of material to create the backing of our sports floors!"*

*For additional **09 62 50** related LEED products, see CSI Section **09 64 00**, along*

with:
http://www.arcat.com/divs/sec/sec096250.shtml

09 62 50 - Leather Flooring

Well, if you have noticed, *CSI Section* ***09 62 50*** has two different headings. <u>Why is this?!</u> Other than in some historical applications, leather flooring, is beautiful, but **is not popular. And, maybe for this reason it is not** currently listed at Arcat.com, but it can be found at 4specs.com. Take a look . . .

- **EcoDomo, Inc.**
  14650 Rothgeb Drive, #F
  Rockville, MD 20850
  Phone: 301.424.7717
  Fax: 301.424.7719

  http://ecodomo.com/floor/

  "Installation of our recycled leather floors in tiles and planks is as easy as carpet tiles, floating floors, or resilient floors. Suitable for light commercial traffic and all non-wet areas of the house."

- **Keleen Leathers**
  1010 Executive Drive, Suite 400
  Westmont, Illinois 60559

  Phone: 630.590.5300
  Fax: 630.590.5213

  http://keleenleathers.com/leather-walls-and-floors/

  "As a family company, we genuinely care about Keleen Leathers and the clients we serve. We pledge to run an honest, open-minded company while continuing to seek amazing new and different products for the interior design and architectural community. (Products are) made and tanned in the USA from Free Range Cows."

For additional **09 62 50** related LEED products, see:
http://www.4specs.com/s/09/09-6250.html

09 63 00 - Brick Flooring

For a selection of **09 63 00** related LEED products, see **Division 04**, along with:
http://www.arcat.com/divs/sec/sec096300.shtml

09 63 40 - Stone Flooring

- **Building Stone Institute (BSI)**
  36 Cougar Lane
  P.O. Box 419
  Chestertown, NY 12817
  Phone: 518.803.4336
  Fax: 518.803.4338

  https://www.marble-institute.com/

- **Marble Institute of America (MIA)**
  380 E. Lorain Street
  Oberlin, OH 44074
  Phone: 440.250.9222
  Fax: 440.774.9222

  https://www.marble-institute.com/

Both the **Building Stone Institute** and the **Marble Institute of America** are united with the name *"The Natural Stone Institute"*. See their website for additional information and services.

For a selection of **09 63 40** related LEED products, see:
http://www.arcat.com/divs/sec/sec096340.shtml

*See CSI Section **04 40 00 - Stone Assemblies,** for related products and information.*

09 64 00 - Wood Flooring

- **Alleghany Mountain Hardwood Flooring**
  501 Main Street
  Emlenton, PA 16373
  Toll Free: 800.867.9441

  http://alleghenymountainhardwoodflooring.com/

"Whether you're a homeowner, a professional installer or an architect, we take the time to answer any questions you may have to make sure you're getting the hardwood flooring that's right for your project."

- **Bruce Hardwood**
  Armstrong Flooring, Inc.
  824 Linden Street
  Bethlehem, PA 18018
  Phone: 610.865.6700

  http://www.bruce.com/flooring

"As the industry's leading North American hardwood flooring manufacturer, Armstrong (Bruce is an Armstrong brand) has a vested interest in protecting the forests where our raw materials come from."

- **OldeWood**

  7557 Willowdale Street
  Magnolia, Ohio 44643
  Toll free: 866.208.WOOD (9663)
  Phone: 866.208.9663
  Fax: 330.866.1735

  http://www.oldewoodltd.com/

  *"What sets **Olde Wood** apart? We Mill It. We Manufacture It. And We Stand Behind It. "*

- **Tiikeri Flooring**

  TorZo Sustainable Surfaces
  2475 Progress Way
  Woodburn, OR 97071
  Toll Free: 866.822.0002

  http://www.torzosurfaces.com/downloads/

  *"**TorZo** ™ manufactures a unique, elegant and sustainable line of pre-finished flooring products made from agricultural byproducts and recycled wood. TorZo infuses these sustainable materials with an acrylic, creating a very hard and durable flooring material - an ideal option for both residential and commercial applications."*

For additional **09 64 00** related LEED products, see:
http://www.arcat.com/divs/sec/sec096400.shtml

*See CSI Section **06 40 00 - Architectural Woodwork,** for related products and information, including reclaimed lumber.*

09 64 69 - Cork Flooring

Cork is a natural "resilient flooring" *(See CSI Section 09 65 00)* and rapidly renewable flooring material.

It is considered to be rapidly renewable - as the cork bark can be cut from the tree approximately every 9 years or so. However, the tree needs to be at least 24 years old. Along with this, cork recycling is becoming more popular as well, be it wine-stoppers or other prior products.

However, since cork grows as a bark on the Cork Oak Tree, commonly grown in the Southwest portion of Europe and into the Northwestern portion of the African Continent, like Bamboo, it technically is not a USA made product.

- **Globus Cork Flooring**

741 E. 136th Street
Bronx, NY 10454
Phone: 718.742.7264
Fax: 718.742.7265

http://www.corkfloor.com/index.html

"A beautiful cork floor is good for your health as well as the environment. We developed our products in the 1990s, launched our factory in 2001 and are the premier US manufacturer of colored cork tiles. We (at Globus) use only renewable wind energy for our manufacturing energy needs since 2007."

- ## USFloors

Corporate Offices
3580 Corporate Dr.
Dalton, GA 30721
Phone: 706.278.9491
Toll Free: 800.404.2675

http://www.usfloorsllc.com/display-category/natural-cork-5/

*"In the fall of 2008, **USFloors** opened domestic floor production facilities in Dalton, GA concentrating on the manufacture of cork, bamboo, and FSC®-Certified hardwood flooring products. The opening of this 100,000 sq. ft. facility, which also includes its corporate offices, makes USFloors the ONLY supplier of cork and bamboo flooring with manufacturing facilities in the US."*

For additional **09 64 69** related LEED products, see:

http://www.arcat.com/divs/sec/sec096469.shtml

09 65 00 - Resilient Flooring

As stated in the quote below, and within the *Resilient Floor Covering Institute* listing, the term resilient flooring typically means durable natural and/or man-made products *"that occupy a middle ground between soft floors, (like carpeting), and hard floors, (like stone and hardwood)."* In other words, a resilient floor will provide "give" without cracking. For more information, see:

http://homerenovations.about.com/od/floors/g/resilientfloor.htm

- **Resilient Floor Covering Institute** (RFCI)

115 Broad Street, Ste. 201
La Grange, GA 30240

http://rfci.com/

*"**RFCI** is all about resilient flooring . . . and resilient flooring is all about sustainability, durability, affordability, and style. It encompasses a surprisingly wide variety of hard surface flooring products - from vinyl and linoleum to rubber and cork. RFCI is an industry trade association of leading resilient flooring manufacturers and suppliers of raw materials, additives, and sundry flooring products for the North American market."*

For a selection of **09 65 00** related LEED products, see:
http://www.arcat.com/divs/sec/sec096500.shtml

09 65 40 - Rubber Flooring

- **Ecore Commercial Flooring** (formerly Dodge)
  P.O. Box 4944
  Lancaster, PA 17604
  Toll Free: 877.258.0843

  http://www.ecorecommercialflooring.com/ecosurfaces/ with 3:42 company overview video

  "For almost 20 years, ECOsurfaces recycled rubber flooring has offered a diverse portfolio of flooring patterns that make a distinct design statement without sacrificing the health of the natural environment."

For additional of **09 65 40** related LEED products, see:
http://www.4specs.com/s/09/09-6540.html

*See CSI Section **12 48 50 – Fitness Room Flooring & Mats,** for related products and information.*

09 65 43 - Linoleum Flooring

Linoleum Flooring, continues the resilient floor topic and was first invented in the mid to late 1800's. It was/is the predecessor to the sheet vinyl flooring of today. Sometimes, people mistake one for the other. And, in recent years, linoleum has made a comeback as it is made from natural materials, such as: Linseed Oil, pine rosin, ground cork dust, wood flour and natural fillers such as ground limestone. Typically, the tell tale sign to determine if it is linoleum (and not vinyl) is the common burlap backing. In earlier years, however, it is learned that linoleum flooring had asbestos content, and can still be found in older buildings today.

In new linoleum installations, it is learned that:
- When linoleum is installed on a slab on grade or actually any application, the surface needs to be floated well, otherwise any imperfections will be noticeably embossed through.

- Sheet linoleum (from a roll) does not wear as well as some other resilient materials, and may need to be patched as necessary.

- The seams between sheet linoleum are welded and can be accent features. See: **https://www.youtube.com/watch?v=DRS6EekXM6s**

- Linoleum tiles are also available, allowing for creative patterns. (Remember to center the pattern within the space. *See CSI Sections **09 30 50** and **09 50 00**).*

- **Armstrong World Industries**
  325 Eagle Valley Rd
  Beech Creek, PA 16822
  Phone: 570.962.2106

  http://www.armstrong.com/flooring/products/linoleum

  "Crafted from natural materials, linoleum flooring is a smart choice for homes (and any building) looking to "go green"."

- **Forbo Flooring North America**
  8 Maplewood Drive
  Humboldt Industrial Park
  Hazleton, PA 18202
  Phone: 800.842.7839
  Fax: 570.450.0258

  http://www.forbo.com/flooring/en-us/products/marmoleum/cfctp7

  *"Naturally beautiful, durable, sustainable and healthy! **Forbo Marmoleum** floors are associated with sustainability, durability, high quality and innovative design. Our Marmoleum range includes solutions for virtually any type of application."*

09 66 00 - Terrazzo Flooring

- **National Terrazzo and Mosaic Association**
  PO Box 2605
  Fredericksburg, TX 78624
  Toll Free: 800.323.9736
  Fax: 888.362.2770

  http://ntma.com/

  *"The **National Terrazzo and Mosaic Association**, Inc. is a full service Non-Profit Trade Association headquartered in Fredericksburg, Texas. The Association establishes national standards for all Terrazzo floor and wall systems and provides complete specifications, color plates and general information to architects and designers at no cost."*

- **EnviroGLAS Products, Inc.**
  309 Gold Street
  Garland, TX 75042
  Phone: 972.276.9451
  Fax: 972.276.4736

  http://enviroglasproducts.com/?page_id=23

Terrazzo flooring has a long and rich history that dates back over 1500 years. Terrazzo, from the Italian word for terraces, is one of the original recycled products – created centuries ago by Venetian workers utilizing the waste chips from slab marble processing. Today terrazzo flooring continues to provide the ultimate in low maintenance and durability, typically lasting the life of the building. Combine that with the lower cost and environmental advantages of ***EnviroGlas'*** *crushed glass aggregate and it is easy to see why Terrazzo is undergoing a Renaissance.*

- **Master Terrazzo, Inc.**

  Toll Free: 888.999.6885

http://masterterrazzo.com/

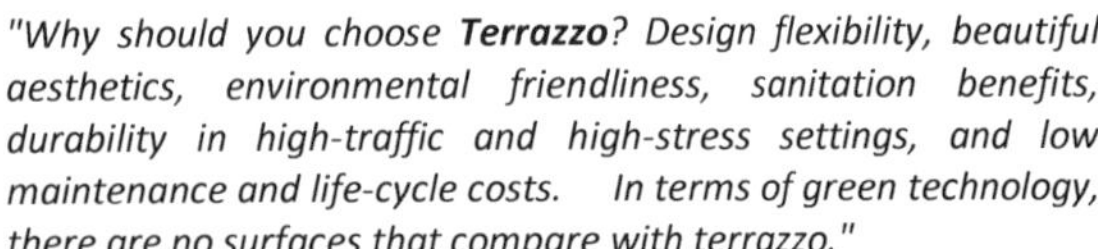

"Why should you choose ***Terrazzo****? Design flexibility, beautiful aesthetics, environmental friendliness, sanitation benefits, durability in high-traffic and high-stress settings, and low maintenance and life-cycle costs. In terms of green technology, there are no surfaces that compare with terrazzo."*

For a selection of **09 66 00** related LEED products, see:
http://www.arcat.com/divs/sec/sec096600.shtml

09 68 00 - Carpeting

Carpeting. We, or at least I have a love - hate relationship with it. Can you relate? Obviously, there are various types and styles. And, as we know, it is pleasant to walk on, and depending on the room and the type of pad and carpet, it can be cozy to lay down on. But, what about all the grime and filth that was tracked in from the outdoors?! This is where not only the Japanese culture is so far ahead, but doing a little research, it also is found that people in Korea, Turkey, Germany, Switzerland and the Scandinavian countries commonly take off their shoes at the door to avoid bringing such *?!?!* into the home, or other buildings.

Carpeting should not be confused with **area rugs** listed in Division 12 - Furnishings. Typically carpeting means it is installed "wall to wall". Wall to wall carpet became popular starting in late 1960's or early 1970's. At this time, in many instances, carpeting was installed over hard wood floors, as a way to modernize these existing homes. Now in recent years, the hardwood floors (commonly installed in most new homes built up to the late 1960's) were covered over and for the most part were protected during these past 50 some years. Now such cherished old growth hardwood is the rage as people are re-recognizing that wall to wall carpeting traps dirt and can have a negative impact on **Indoor Air Quality** (IAQ). With this, rugs and walk off mats at entrances are critical.

Some of the more revolutionary breakthroughs in the carpet industry has been:

- The development of **Anso Nylon** carpet fibers. *(Thanks to Shaw Floors, all carpet made from this Type 6 nylon fiber can be recycled, making it Cradle to Cradle Silver certified!)* ***http://www.c2ccertified.org/products/scorecard/anso_and_unbranded_type_6_nylon_residential_carpet***

- The development of **carpet tiles**. *(Thanks to Interface Carpet Founder and innovator, Ray Anderson in 1994 - the industry has been changed forever!)* ***http://www.interfaceglobal.com/sustainability.aspx***

*- See CSI Section **12 48 00 – Rugs and Mats,** for related products and information.*

09 68 13 - Tile Carpeting

Carpet Tiles / Carpet Squares have the benefits of ease of installation and the ability to replace tiles if and when needed. Some use adhesive and others use corner tack strips that connect the carpet together, but it floats on the floor surface. And like other floor tiles, these are typically centered within the space, for design and installation reasons. The size of the tiles vary by manufacturer.

- **FLOR**
  600 W. Van Buren Street, Suite 800
  Chicago, IL 60607
  Phone: 866.952.4093

  http://www.flor.com/

  *"**FLOR** products are constructed of thoughtfully chosen materials that provide superior durability, stain resistance and colorfastness under typical living and environmental conditions."*

- **Interface Americas, Inc.**
  1503 Orchard Hill Road
  LaGrange, GA 30241
  Toll Free: 800.634.6032

  http://www.interface.com/US/en-US/global

  "We make carpet tile, but we sell design. Timeless. Beautiful. Enduring. And, refreshingly new."

  Interface is the originators of carpet tile.

- **StaticStop Flooring**

A Division of SelecTech, Inc.
33 Wales Avenue, Suite F
Avon, MA 02322
Phone: 508.583.3200
Fax: 877.738.4537

http://www.staticstop.com/

Static Stop Flooring provides a variety of carpet tile and resilient tile products that offer Electrostatic Discharge (ESD) properties. *"Both our modular and glue-down flooring are certified MAS Green. Only products that pass the stringent MAS tests can display the MAS Certified Green logo. This logo is your assurance our flooring is certified for low VOC emissions."*

For additional **09 68 13** related LEED products, see:
http://www.arcat.com/divs/sec/sec096813.shtml

09 68 16 - Sheet Carpeting (Roll)

Sheet carpeting typically comes in 6 foot and 12 foot roll widths. Seams are carefully thought out and disguised by installers whether Cut Pile, Loop Pile, Multi-level Loop Pile or Cut Loop Pile carpet is specified and installed, while using any of the various carpet fibers available. See a related illustrated guide at:
http://www.carpet-rug.org/Carpet-for-the-Home/Selecting-the-Right-Carpet/Carpet-and-Rug-Construction.aspx

- **The Carpet and Rug Institute (CRI)**

Street Address:
100 South Hamilton Street
Dalton, GA 30720

Mailing Address:
P.O. Box 2048
Dalton, GA 30722-2048

Phone: 706.278.3176
Fax: 706.278.8835

http://www.carpet-rug.org/

"Today, indoor air quality (IAQ) is an important environmental consideration, especially since we spend approximately 90 percent of our time indoors. In 1992, ***CRI*** *launched its Green Label program to test carpet, cushions and adhesives to help specifiers identify products with very low emissions of Volatile Organic Compounds (VOCs)."*

For a selection of **09 68 16** related LEED products, see:
http://www.arcat.com/divs/sec/sec096816.shtml

09 69 00 - Access Flooring

Access Flooring, or sometimes known as **Raised Access Flooring** offer the opportunity to run utilities below the flooring, providing low cost office space flexibility. Floor tiles are generally 24 inch square and can be designed for various plenum heights, however a 12" air plenum is common. For more information see:

http://www.irvineaccessfloors.com/access-floors/why-access-floors/ with 2:00 animated video.

- **Irvine Access Floors**
  9425 Washington Blvd., Suite Y-WW
  Laurel, MD 20723
  Toll Free: 800.969.8870
  Phone:301.617.9333
  Fax:301.617.9907

  http://www.irvineaccessfloors.com/

  "Trust the expertise of an industry leader. ***Irvine Access Floors*** *delivers American-made access floor systems on-time and on-budget."*

- 
  Tate, Inc.
  Corporate Headquarters
  7510 Montevideo Road
  Jessup, MD 20794
  Toll Free: 800.231.7788
  Fax: 410.799.4207

  http://tateinc.com/commercial/access-floor-systems

  "To be successful in the global economy, companies need to invest properly, adapt to changes in technology, create a healthy and comfortable work environment for attracting and keeping employees, and operate in an environmentally responsible manner. Tate's commitment to SustainAbility includes goals and actions for continuous improvement in the areas of the Environment, Customers, Community, and Company."

For additional **09 69 00** related LEED products, see:

http://www.4specs.com/s/09/09-6900.html

09 70 00 - Wall Finishes

Wall Finishes addresses various wall coverings, such as textile wall finishes, carpet, cork, vinyl, wall papers, tile, and other items that are specified and applied to walls. Unfortunately, it is the wall finish items (that are adhered to the wall substrate) that typically ware out - either from normal wear or by **perceived obsolescence**.

One thing that I am fascinated however is how various trends repeat themselves. For example, taking a tour in a historic home, it was learned how reflective wall paper was used in the late 1800's as a way to reflect light in rooms. And, today with the emphasis and importance of daylighting, it is interesting to find similar wall paper that has come back in vogue. See:
https://www.houzz.com/reflective-wallpaper

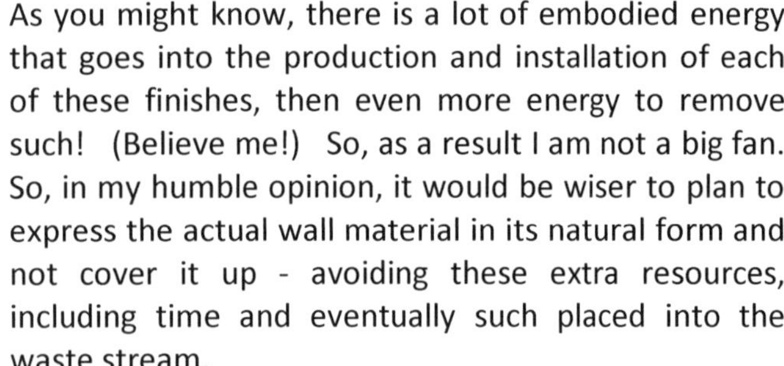

As you might know, there is a lot of embodied energy that goes into the production and installation of each of these finishes, then even more energy to remove such! (Believe me!) So, as a result I am not a big fan. So, in my humble opinion, it would be wiser to plan to express the actual wall material in its natural form and not cover it up - avoiding these extra resources, including time and eventually such placed into the waste stream.

But, you may disagree. So, take a look . . .

- **Sensitile Systems**
  1735 Holmes Rd.
  Ypsilanti, MI, 48198
  Phone: 313.872.6314

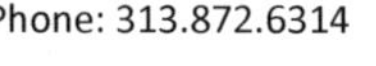

http://sensitile.com/products

*"**Sensitile Systems** is an internationally recognized manufacturer of inspiring and sustainable products for a wide range of architectural and design applications. Our award winning, patented products are manufactured exclusively in Michigan, USA using cutting-edge technologies and green manufacturing processes and have been used in landmark projects worldwide. Each material line has a unique aesthetic that dynamically integrates and interacts with different environments and its inhabitants to create to create unforgettable experiences."*

*See CSI Section **09 30 50*** along with the below website for ***09 70 00*** related LEED products:
http://www.arcat.com/divs/sec/sec097000.shtml

09 80 00 - Acoustic Treatment

- **SoundPly** by Navy Island, Inc.
  275 Marie Ave. East,
  West St. Paul, MN 55118
  Phone: 651.451.4454

Fax: 651.451.4484

http://www.soundply.com/

Navy Island offers a variety of crafted products, including: Wood Acoustic Ceiling panels, planks and baffle systems; Wood Acoustic Wall Panels; Fire Rated Wood Door Frames and Lites. Architectural Wood Veneers; and Flush Wood Interior Doors.

- **Tectum, Inc.**
  PO Box 3002
  Newark, OH 43058
  Toll Free: 888.977.9691
  Phone: 740.345.9691

http://www.tectum.com/

*"**Acoustics** are a key element of good building design and have been proven to affect an occupant's perception of the quality of a space. No matter what kind of space, good acoustics make the sounds cleaner, clearer and more easily understood."*

- **dB Sound Control Systems**
  United Plastics Corporation
  511 Hay Street
  Mt. Airy, NC 27030
  Phone: 336.786.2127
  Fax: 336.786.6966

http://www.dbsoundcontrol.com/home

*"**MyTheater® Professional Acoustic Panels** improve media room and home theater acoustics by absorbing echoes to create higher-quality sound"*

*See also CSI Section **09 50 00 - Ceilings** for more information on acoustics*

For additional **09 80 00** related LEED products, see:
http://www.arcat.com/divs/sec/sec098000.shtml

09 90 00 - Painting and Coating

There is so much more to paints and coatings than to simply brush or spray it onto a surface. The below bulleted information is a start, but also there is the study of color as related to the **Color Wheel** and all the related theories and psychology. See:
https://www.colormatters.com/color-and-design/basic-color-theory

- **Exterior Paint Problems on Historic Woodwork**
  US Department of the Interior
  Technical Preservation Services
  Preservation Brief 10

https://www.nps.gov/tps/how-to-preserve/briefs.htm

- ## Painting Historic Interiors

US Department of the Interior
Technical Preservation Services
Preservation Brief 28

https://www.nps.gov/tps/how-to-preserve/briefs.htm

- ## Breathable Mineral Paints & Stains

Limeworks.us
3145 State Road
Telford, PA 18969
Phone: 215.536.6706
Fax: 215.453.1310

http://www.limeworks.us/home.php

And, specifically the breathable silicate paints and stain products can be found here:
http://www.limeworks.us/blog/uncategorized/mineral-paints-and-stains-make-a-breathable-and-durable-mural-on-a-historic-building/

*"**LimeWorks.us** is a full-service supply company committed to providing the resources you need for performing professional masonry restoration and to build sustainable structures."*

- ## Devine Color Paint

333 S. State Street
Lake Oswego, OR 97034
Phone: 503.387.5840

https://www.devinecolor.com/experience-paint-illuminated/the-devine-difference with 3:07 color based video

*"**Devine Color** is made differently. Our exclusive, color-illuminating formula is infused with premium-quality pigments that embrace light and let true, rich, radiant color shine through. Devine Color is luxuriously creamy and coats walls abundantly with a soft, lush finish that's durable and washable. It's low odor, zero VOC (as calculated by EPA method 24).*

- ## Roma Domus Line Mineral Paints

ROMABIO
554 North Avenue NW, Ste. B
Atlanta, GA 30318
Phone: 678.905.3700

http://romabio.com/

*"**ROMABIO** manufactures mineral paints, plasters, limewash products, and more. Mineral paints are the healthy alternative to acrylic paints, completely Zero VOC, toxin-free, odorless, asthma-free, and naturally mold-resistant. Different than the acrylic based Zero VOC paints in the industry, ROMABIO mineral paint is made from natural, raw materials manufactured without toxic chemicals that cause Sick Building Syndrome, cancer or asthma."*

ROMA Domus Mineral Paints placed 3rd in the Cradle to Cradle product innovation challenge:

http://inhabitat.com/meet-the-winners-of-the-first-cradle-to-cradle-product-innovation-challenge/

- ## The Old Fashioned Milk Paint Co., Inc.

  436 Main Street
  Groton, MA 01450
  Toll Free Order Line: 866.350.6455
  Customer Service: 978.448.6336
  Fax: 978.448.2754

  http://www.milkpaint.com/about.html

  "Our milk paint is made today with the same basic ingredients used for the past hundreds of years- milk protein, lime and earth pigments. The look and feel of a surface painted today with our milk paint is no different than what was found on furniture, walls and floors in country houses in Colonial America."

*- See CSI Section **12 11 00 – Painted and Photo Murals,** for related products and information.*

09 96 00 - High-Performance Coatings

- ## Aarmoutherm

  2009 Mount Vernon Avenue
  Riverside, CA 92507

  Phone: 951.787.8503

  http://www.aarmourtherm.net/

  Aarmoutherm manufactures edible insulation and paint coatings for the home, industry and other applications. See related information and videos at above website.

- ## Hy-Tech Thermal Solutions

  P.O. Box 216,
  Melbourne, FL 32902
  Phone: 321.984.9777
  Fax: 321.984.1022

  http://hytechsales.com/insulating_paint_additives.html

  Hy-Tech manufactures insulating paint and industrial coatings for the home, industry and transportation.

- ## Insuladd Environmental LTD

  PO Box 260106
  Plano, TX 75026
  Toll Free: 888.813.5065
  Phone: 214.738.1609

  http://www.insuladd.com/

  *"**Insuladd Energy Saving Paint Additive**, which is made up of a complex blend of ceramic microspheres, is designed to mix with any type of paint, including latex house paint, industrial coating, roof coating, epoxy, urethane, and even high temperature paint. Once it has been mixed and applied to the walls and ceilings of a home, it acts as a thermal radiant barrier."*

- **Super Therm**

  Superior Coating Solutions
  1641 3rd Avenue Apartment 14a
  New York, NY 10128
  Phone: 917.836.7816

  http://www.superiorcoatingsolutions.com/

  Bob Vila: Super Therm® Ceramic Insulation
  https://www.youtube.com/watch?v=LtBICM4F1Gs 2:17
  plus other video testimonials at Superior Coatings website

  But, there is another side to the story. See:
  http://www.greenbuildingadvisor.com/blogs/dept/musings/insulating-paint-salesman-tripped-his-own-product

It should be noted that no matter what finish is being applied, ***proper surface preparation*** *is critical to the longevity of applied finishes. Make sure that proper surface preparation is specified and adhered to. (No 'pun' intended).* ☺

End of Division 09

Space for notes:

DIVISION 10 - Specialties

When it comes to ***Specialties,*** it appears at least to me, that this Division is a 'catch all' of very important items. Take a look. Some very cool stuff. Yet, the only way that it appears that they relate to each other is that they all can be specified for the construction of a building. ☺

10 11 00 - Visual Display Units

This section includes Chalkboards; Markerboards; Electronic (Smart) Boards; and Tackboards / Corkboards.

For a selection of **10 11 00** related LEED products, see:

http://www.arcat.com/divs/sec/sec101100.shtml

10 12 00 - Display Cases

For a selection of **10 12 00** related LEED products, see:

http://www.arcat.com/divs/sec/sec101200.shtml

10 14 00 - Signage

This section includes exterior and interior signs that are so necessary to direct outside and inside the buildings we design and occupy. With this it is important to consider exterior sign graphics, letter size and font for visibility and sign heights to address potential zoning and potential Historic regulatory requirements. See the **United States Sign Council** *Sign Legibility Rules of Thumb* document provided below:
http://www.usscfoundation.org/USSCSignLegiRulesThumb.pdf

The Preservation of Historic Signs

National Park Service
US Department of the Interior
Technical Preservation Services
Preservation Brief 25

https://www.nps.gov/tps/how-to-preserve/briefs.htm

Then continuing, there are of course the interior signs which include the actual sign design, materials used, text height, Braille, and the Universal Design based mounting heights that should accommodate *all* people. See **ADA Standards for Accessible Design**, *Section **703 - Signage***:

https://www.ada.gov/regs2010/2010ADAStandards/2010ADAstandards.htm#c7

All exterior and interior signs should provide a positive "**wayfinding**" experience, guiding *everyone* well, while using environmentally friendly products.

For a selection of **10 14 00** related LEED products, see:
http://www.arcat.com/divs/sec/sec101400.shtml

10 21 13 - Toilet Compartments

The older I get, the more I despise public restrooms. They are so necessary, yet so undesirable. *Dwell* has captured what the great folks at *Smart Design* has achieved. *The Bathroom Reinvented: Universal Design in Public Bathrooms*
https://www.youtube.com/watch?v=Ymhn2ulHosg
7:00

Then, as a follow up to the above, *The Bathroom Reinvented: Universal Design - Part II* (at Madison Square Park, NYC)
https://www.youtube.com/watch?v=1eylxXbUB_0
5:38

Every once in a while though, you, like I will find a business or even a highway Rest Stop that is just immaculate while using the typical fixtures and stall configurations. And, typically when cared for, it is interesting to see that the patrons also follow suit in taking care of it, as they notice and respect the cleanliness. To this I thank all involved!

In a multi-stall restroom, one way to help with the maintenance is to recognize that there is a choice in the actual toilet partition materials, such as, depending on the location, and if prone to vandalism, the Solid Plastic (HDPE) partitions are commonly best. Here is one example of this:

https://www.bradleycorp.com/applicationguides/plastic-partitions

Other partition materials typically are: Powder Coated Steel; Plastic Laminate; Solid Phenolic Core; and Stainless Steel. This link describes each of these characteristics. http://bencoinc.com/announcements/toilet-partition-material-basics

Then, for ease of floor cleaning, the ceiling hung partitions are best, eliminating the need to mop around the partition legs. However, such design can be more susceptible to be hung on - as some individuals may attempt to vandalize these by pulling them down from the ceiling mounts. This link shows the different mounting configurations and provides additional toilet partition material data and illustrations.
http://www.americanspecialties.com/catalogdocs/AccurateCat2013.pdf

With this section on Toilet Compartments and public restroom thoughts in general, please be mindful that the International Building Code (IBC) dictates the minimum number of toilets, lavatory sinks, potential substitution for urinals, along with mop sinks and drinking fountains. See the following (IBC) Section **2902 - *Minimum Plumbing Facilities*** I-Codes website, for more information.

So to also include the Residential requirements for *One and Two Family Dwellings*, the International Residential Code (IRC) *Section **R307 - Toilet, Bath, and Shower Spaces*** describes related requirements.

Use website to access the *I-Codes, PremiumACCESS or the Free PublicACCESS*: **https://codes.iccsafe.org/**

For **Universal Design** requirements, see the wonderful **Bobrick Planning Guide for Accessible Restrooms**:
http://www.bobrick.com/documents/planningguide.pdf

*See **Division 22 - Plumbing,** for related products and information .*

This section commonly focuses also on **Toilet Partitions**. These partitions have not changed much over the years, nor has the abuse that public restrooms face. The product material choices today are typically Powder Coated Steel, Stainless Steel, Plastic Laminate, Solid Plastic, Color-Thru Phenolic, and a few others. You may come across some historic ones made from slate or marble potentially still in use, or salvaged. And, you might come across ones made from wood, with a clear or painted finish. However, it is unfortunate that graffiti, vandalism and overall maintenance end up being the main challenges for any of the choices available.

So, along with the Toilet Partition materials choices, how the partitions are actually supported should also be a consideration. When considering the end users could influence how the partition is supported or how the room is cleaned. The ceiling mounted style helps dramatically with mopping, since the partition legs/supports do not touch the floor. However, from a vandalism point of view, these would need to be well secured to the ceiling to prevent someone from pulling them off the ceiling. (Just some of all the important details to consider.) ☺

For a selection of **10 22 00** related LEED products, see:
http://www.arcat.com/divs/sec/sec102200.shtml

10 22 23 - Portable Partitions, Screens and Panels

The concept of on operable room divider type partitions, that can offer the flexibility to divide a larger space into smaller spaces for various events and activities. This, as you know, is a sustainable design concept by itself!

Portable Partitions (or **Demountable Partitions**) allow a semi-permanent wall to be removed and reused in a new configuration elsewhere, while not disturbing the floor finish or acoustic ceiling tile. Typically more expensive initially than gypsum board and metal studs, however there is no dust, and such systems can provide significant daylighting (**borrowed lite**) opportunities.

- **nxtWall**
  5200 South Sprinkle Road
  Kalamazoo, MI 49002
  Phone: 269.488.2752

  http://www.nxtwall.com/

"There's a reason why people say... "it's a true facilities manager's product" ... we make moveable walls that are a real solution. ***NxtWall*** *understands who is left moving and adapting these walls. Three very important aspects are designed into all our walls: Simplicity, Flexibility, and Sustainability."*

For additional **10 22 23** related LEED Products, see:
http://www.arcat.com/divs/sec/sec102223.shtml

10 25 00 - Service Walls

Service Walls are commonly those that are along the bed 'head' or 'service' wall in a hospital patients' room, providing the necessary equipment for care and monitoring.

For a selection of **10 25 00** related LEED products, see:
http://www.arcat.com/divs/sec/sec102500.shtml

*- See CSI Section **11 70 00 - Health Care Equipment,** for related products and information.*

10 26 00 - Wall and Door Protection

Wall and Door Protection has come a long way in what is available and also the sustainable products that are being used to create such.

Often found in institutional environments such as nursing homes and hospitals, many of the wall protection products double as gurney bumpers and handrails along the corridors. Other wall and door protection products guard corners of walls and door frames from damage for unlimited building types and uses.

See the great team at *Construction Specialties* for their innovative products, including *Acrovyn 4000*, a Cradle to Cradle (C2C) product that offers manufacturing transparency.

- **Construction Specialties** (C/S)
  6696 State Route 405
  Muncy, PA 17756
  Toll Free: 800.233.8493
  Phone: 570.546.5941

  http://www.c-sgroup.com/acrovyn/wall-protection

  With various locations and various products, per their website, this **C/S** location addresses: "*Acrovyn, Entrance Flooring & Expansion Joint Covers*".

For additional **10 26 00** related LEED products, see:
http://www.arcat.com/divs/sec/sec102600.shtml

10 26 41 - Bullet Resistant Panels

The protection of life is the most sustainable provision one can provide a client. What can you do to incorporate such panels into the projects that you design? And, which of your clients needs to be protected in this way?

For a selection of **10 26 41** related LEED products, see:
http://www.arcat.com/divs/sec/sec102641.shtml

10 28 00 - Toilet, Bath, and Laundry Accessories

This section includes the hand dryers, toilet paper, paper towel, soap and any other dispensers, grab bars, baby changing stations, and other common accessories found within a restroom / bathroom.

Along with the accessories comes the necessary mounting heights for such. See this and other information in the ***Bobrick Planning Guide for Accessible Restrooms***:
http://www.bobrick.com/documents/planningguide.pdf

When it comes to paper towels vs. hand dryers, what is your preference? There is the mindset that a hand dryer needs less maintenance (no refilling of paper towels, less garbage and no need to replace batteries), but then what about the piercing noise that most of these supersonic hand dryers make?! And, what about the electricity used? Or the initial expense for some of the more elaborate models? (Nice, but these could really break your budget or you could buy a lot of paper towels instead.)

- Dyson Airblade cost:
http://www.handdryer.com/dyson-airblade-hand-dryers/

- Georgia Pacific enMotion cost:
https://www.amazon.com/59462-Automated-Touchless-Dispenser-Translucent/dp/B015X6E7FY

When it comes to the paper towels, there is something nice about having it in your hand so to grasp the door hardware with, then toss it as you leave. But, if a public restroom door was designed and installed to opened outward (so to not swing into the restroom), this would also help solve this problem. Details. Details. Details! Choices. Choices. Choices!

Whichever you prefer, make sure your client agrees. And, most importantly, wash your hands!

- **Dyson Inc.**
  600 W. Chicago Ave, Suite 275
  Chicago, IL 60654
  Phone: 844.679.1647

http://www.dyson.com/hand-dryers.aspx

*"**Dyson Airblade™** hand dryers can cut your carbon footprint. They produce up to 79% less CO_2 than other hand dryers and up to 76% less CO_2 than recycled paper towels[4]."*

- **Georgia-Pacific Professional**
  133 Peachtree St NE
  Atlanta, GA, 30303
  Phone: 866.435.5647

http://www.gppro.com/category_towel.aspx

(The enMotion is a) *"Touchless one-at-a-time dispensing helps minimize cross-contamination risk and creates a more hygienic environment. Not to mention, it reduces usage by up to 30% compared to standard folded towels."*

For additional **10 28 00** related LEED products, see:
http://www.arcat.com/divs/sec/sec102800.shtml

*- See CSI Section **10 21 13 - Toilet Compartments,** and **Division 22 - Plumbing,** for related products and information.*

10 28 19 - Tub and Shower Doors

Now this is one of those topics that you would think that because it is a door it would be in Division 08 - Openings. However, since it is a specialty to bathrooms, such glass doors belong here.

For a selection of **10 28 19** related LEED products, see:
http://www.arcat.com/divs/sec/sec102819.shtml

10 28 23 - Laundry Accessories

If you have not already noticed, and as mentioned at the beginning of Division 10, this Division is quite the 'catch all' of some necessary items including items related to laundry. One of the more practical, space saving products are the Built-in Ironing Boards.

- **Iron-A-Way, Inc.**
  220 W. Jackson St.
  Morton, IL 61550
  Toll Free: 800.536.9495
  Phone: 309.266.7232
  Fax: 309.266.5088
  http://ironaway.com/

"When it comes to concern for your ironing needs, any constraints on square footage will be improved with the installation of a built in or wall mounted ironing center. The ironing center will not only free up floor space, but its functionality is neatly tucked out of sight until it's in use. It will become a convenience you are sure to adapt too quickly."

This topic brings to mind one of the *Three Stooges* video clips, showing *Shemp vs. an Ironing Board.* ☺
https://www.youtube.com/watch?v=4ZG6l_wF4e0 1:28

10 31 00 - Manufactured Fireplaces

Any kind of fireplace brings to mind cold evenings where the family might be gathered to enjoy the ambiance and the heat from such.

Manufactured Fireplaces, or sometimes known as pre-fabricated or Zero-Clearance fireplaces are those that can be installed almost anywhere. They are made from insulated steel, so technically it can be installed next to wood framing with unlimited finishes. Since they are not solid masonry, they do not have the weight of such, thus they do not require a foundation to support it, thus less cost.

This section includes the various options from typical wood log burning and natural fuel gas units. This section also includes Outdoor Fireplaces.

For a selection of **10 30 00** related LEED products, see:
http://www.arcat.com/divs/sec/sec103000.shtml

10 32 00 - Fireplace Specialties

This section includes the energy saving dampers, glass doors, fireplace water heaters, fireplace Inserts (that are placed into an existing solid-masonry fireplace firebox), and fireplace screens.

It was hoped to list the innovative Lyemance Top Mount Damper and its manufacturer, however, no contact information is found, only its distributors. Why?! What is found however is the company overview, stating, *"Lymance International, Inc. manufacturers fireplace dampers to reduce heat loss and seal out sleet, snow, rain, birds, bats, and insects. As of 1997, Lymance International, Inc. operates as a subsidiary of Copperfield Chimney Supply, Inc. "*
http://www.bloomberg.com/research/stocks/private/snapshot.asp?privcapId=3116387

*See **Lyemance Damper:***
https://www.youtube.com/watch?v=xAu3Rs3wwSA 4:28

- **Stoll Fireplace, Inc.**
  153 Hwy 201
  Abbeville, SC 29620
  Toll Free: 800.421.0771
  Fax: 864.446.2172

  http://www.stollfireplace.com/general/home

  *"For over 40 years **Stoll** metalsmiths have been building some of the industry's finest and most elegant fireplace doors, screens and accessories. Our product line continues to evolve, but our dedication to quality and customer service remain as strong as ever."*

- **Woodland Direct, Inc.**
  13287 W. Star Drive
  Shelby Charter Township, MI 48315
  Toll Free: 800.919.1904

  http://www.woodlanddirect.com/Fireplace-

Accessories/Made-in-USA-Fireplace-Doors

*"**Woodland Direct** is one of the largest e-commerce companies focused on the Fireplace, Chimney, Wood Stove, and Outdoor Living markets. Our founders have come from a Fireplace & Chimney background that reaches back to over 25 years of knowledge and experience."*

10 35 00 - Stoves

Stoves should not be confused with a stove appliance found commonly in a kitchen for cooking. This stove would be a pre-manufactured heating device using either logs, biomass pellets (made from sawdust or agricultural-waste), coal, corn, natural gas or electricity (with an artificial flame).

- **Vermont Castings**
  20 Constitution Boulevard South
  Shelton, CT 06484
  Phone: 855.415.0071

  http://www.vermontcastings.com/

*"For 35 years we've been committed to making eco-friendly products with fine craftsmanship. No other manufacturer can match **Vermont Castings'** history of consistently exceeding the EPA's standards for emissions."*

For additional **10 35 00** related LEED products, see:
http://www.arcat.com/divs/sec/sec103500.shtml

10 44 00 - Fire Protection Specialties

With all the charm and elegance of an indoor Fireplace or Stove that does not have an artificial flame, also comes the reality of fire potential and Carbon Monoxide poisoning! See the following International Residential Code (IRC) sections:

- Section ***R313 - Automatic Fire Sprinkler Systems***

- Section ***R314 - Smoke Alarms***

- Section ***R315 - Carbon Monoxide Alarms***

Use website to access the *I-Codes, PremiumACCESS or the Free PublicACCESS*: **https://codes.iccsafe.org/**

CSI Section ***10 44 00*** includes **Fire Extinguisher Cabinets** and **Fire Blankets.** It should be noted that Fire Extinguisher Cabinets can look attractive with a bronze finish and tinted glass. Nice enough where it could be recessed into walls in a residence. See:

- **Larsen's Manufacturing Co.**
  Florida Division
  3130 N.W. 17th St.
  Ft. Lauderdale, FL 33311
  Toll Free: 800.262.3473
  Phone: 954.486.3325
  Fax: 954.486.3352

  Minnesota Division
  7421 Commerce Lane N.E.
  Minneapolis, MN 55432
  Toll Free: 800.527.7367
  Phone: 763.571.1181
  Fax: 763.571.6900

  http://www.larsensmfg.com/fire_extinguishers/fire_extinguisher_cabinets_cameo_series.html

For a selection of **10 44 00** related LEED products, see:
http://www.arcat.com/divs/sec/sec104400.shtml

See ***Division 21 - Fire Suppression,*** *for other related products.*

10 51 00 - Lockers

Lockers focuses on the various materials commonly used to manufacture lockers, including wood, metal, plastic laminate-clad lockers, plastic, recycled plastic, and glass.

For a selection of **10 51 00** related LEED products, see:
http://www.arcat.com/divs/sec/sec105100.shtml

10 55 00 - Postal Specialties

This section addresses individual and centralized mail delivery boxes, such as the P.O. Box type provided and found in a Post Office, including those common in an apartment complex, dorm, or even some recent housing communities.

It also includes various other mail boxes as well as mail chutes.

For a selection of **10 55 00** related LEED products, see:
http://www.arcat.com/divs/sec/sec105500.shtml

10 56 00 - Storage Shelving

Storage Shelving includes Mobile Storage Shelving Units (that offer on-site high density storage solutions); Industrial Storage Shelving; and Library Shelving.

- **Brodart Contract Furniture Division**
  Clinton County Industrial Park
  280 North Road
  McElhattan, PA 17748-0280
  Toll-Free: 888.521.1884
  Phone: 570.769.7412
  Fax: 570.769.7641

  http://www.brodartfurniture.com/

  *"**Brodart Company** guarantees that each piece of furniture bearing our name is designed to the exact requirements of our customers. By using the highest quality of materials, our products are proudly manufactured in the USA and are built to last. Brodart joins the building industry to promote environmentally responsible building practices."*

- **Datum Storage Solutions**
  89 Church Road
  PO Box 355
  Emigsville PA 17318
  Toll Free: 800.828.8018

  http://www.datumstorage.com/

  *Since 1968, **Datum®** has been a leader in providing storage solutions for commercial, industrial and government use. More than 40,000 companies covering every industry have relied on Datum to design, manufacture and install high quality storage solutions, including mobile storage systems, filing systems, workstations, computer security equipment and more.*

10 70 00 - Exterior Protection

This section includes Awnings, Canopies, Car Shelters, Walkway Coverings; and Transportation Stop Shelters

- **The Use of Awnings on Historic Buildings: Repair, Replacement and New Design**
  National Park Service
  US Department of the Interior

Technical Preservation Services

Preservation Brief 44

https://www.nps.gov/tps/how-to-preserve/briefs.htm

- **MASA Architectural Awnings**

  250 Stelton Rd. , Suite 1
  Piscataway, NJ 08854
  Phone: 732.453.6120 Ext. 123
  Fax: 732.453.6126

  http://www.architecturalcanopies.com/

  *"Here's your opportunity to make your unique design vision a reality. If you can dream it, we can design, build, and install it! **MASA** designs can be based on completely custom ideas or a rethought version of one of our standard products. Many are custom, welded aluminum structures that defy convention."*

10 71 13 - Exterior Sun Control Devices

- **Construction Specialties (C/S)**

  49 Meeker Avenue
  Cranford, NJ 07016
  Toll Free: 800.631.7379
  Phone: 908.272.5200

  http://www.c-sgroup.com/sun-controls

  With various locations and various products, per their website, this **C/S** location addresses: "*Louvers, Grilles & Sun Controls*".

  *"**C/S**, the leader in Sun Controls and architectural Louvers, has joined forces with CENTRIA, the leader in metal wall systems, to offer specifiers and owners a totally integrated facade engineered to work together perfectly."*

See CENTRIA related information within *CSI Section* ***07 42 63*** - ***Fabricated Wall Panel Assemblies***

For additional **10 71 13** related LEED products, see:

http://www.arcat.com/divs/sec/sec107113.shtml

10 71 16 - Storm Panels

Storms can be devastating. When designing, it is critical to select products and integrate them into the design for longevity. Select such products that would stand up to tornados and hurricanes. One solution is to incorporate storm panels (or window glass / glazing safety film) for glass doors and window protection.

- **AMF Building Products**

1501 53rd Street
West Palm Beach, FL 33407
Phone: 561.790.5799
Fax: 561.790.2320

http://www.amfbp.com/index.php/shutters

AMF manufactures various shutters, including the *"Aluminum Storm panels (which) are roll-formed from high-strength tempered aluminium alloy coil stock. They are easy to install, lightweight and can be stored compactly. The panels can be installed using our header, f-track or stud angle mounting systems and our full line of hurricane hardware items."*

*- See CSI Section **08 87 13 - Window Film,** for related products and information .*

10 74 00 - Manufactured Exterior Specialties

This section includes **Exterior Clocks, Cupolas, Spires, Steeples, Weathervanes**, and **Window Wells**.

Cupolas are an original way to bring light in and circulate air. It provided the original **"Stack Effect"** prior to when central air conditioning became popular in the 1970's. See "***Cooling Your House With Free Air***", Popular Science, July 1963 - Pg 99-101.

Not only are Cupolas decorative, but functional as well.

- **Mailbox Shoppe**
  PO Box 618
  Centerport, NY 11721
  Toll Free: 800.330.3309

  http://www.mailboxnet.com/cupolas.html

  The **Mailbox Shoppe** manufactures various products including custom made, ventable Cupolas to order.

  "Ventilation is the primary function of a cupola. Most cupolas have adjustable vents to control airflow into attic spaces. Cupolas are also used to bring more light into a room and increase the sense of internal space. Some cupolas are built large enough to walk into and offer a high vantage point."

Cupolas: *Some additional information*

http://www.greenbuildingadvisor.com/blogs/dept/design-matters/passive-ventilation-and-natural-light-cupola

- **ScapeWEL® Window Wells**
  The Bilco Company
  Corporate Headquarters
  P.O. Box 1203
  New Haven, CT 06505
  Phone: 203.934.6363
  Fax: 203.535.1582

Manufacturing
536 Highway 463 South
Truman, AR 72472
Phone: 870.483.5118
Fax: 870.483.7870

Manufacturing
3400 Jim Grainger Road
Zanesville, OH 43701
Phone: 740.455.9020
Fax: 740.455.9030

http://www.bilco.com/foundations/store/storepage.asp?page=ww

(ScapeWEL) *adds natural daylight and meets building code requirements for emergency egress in finished basement areas. Innovative step design aids egress and can be landscaped with plants or flowers for added visual enhancement.*

Emergency Egress: *Some additional information*

- See International Residential Code (IRC), *Section **R310 - Emergency Escape and Rescue Openings*** for code requirements.

Use website to access the *I-Codes, PremiumACCESS or the Free PublicACCESS*: **https://codes.iccsafe.org/**

10 75 00 - Flag Poles

The US Flag is a symbol of freedom! And, this freedom did <u>not</u> come free, but it was the result of many battles, wars and the ultimate sacrifice that was made by those serving our Country. Sometimes, we might take these freedoms for granite, but once taken away, it would be a loss that would be hard to reinstate. If you ever have an opportunity to visit another country and learn about its ways, you may be very thankful when you return, that you have the privilege to call at least one of these 50 States, (that have been united under one flag), *your* home.

To display the flag, recognize that there is proper flag etiquette, or otherwise known as ***The Flag Code,*** to show respect and honor. **http://www.usflag.org/uscode36.html#USFC**

If not know, there also is ***Flag Etiquette*** as communicated below, as found at: **http://www.usflag.org/flagetiquette.html**

- The flag should never be dipped to any person or

thing. It is flown upside down only as a distress signal.

- The flag should not be used as a drapery, or for covering a speakers desk, draping a platform, or for any decoration in general. Bunting of blue, white and red stripes is available for these purposes. The blue stripe of the bunting shold be on the top.
- The flag should never be use for any advertising purpose. It should not be embroidered, printed or otherwise impressed on such articles as cushions, handkerchiefs, napkins, boxes, or anything intended to be discarded after temporary use. Advertising signs should not be attached to the staff or halyard.
- The flag should not be used as part of a costume or athletic uniform, except that a flag patch may be used on the uniform of military personnel, fireman, policeman and members of patriotic organizations.
- The flag should never have placed on it, or attached to it, any marks, insignia, letter, word, number, figure, or drawing of any kind.
- The flag should never be used as a receptacle for receiving, holding, carrying, or delivering anything.
- When the flag is lowered, no part of it should touch the ground or any other object; it should be received by waiting hands and arms. To store the flag it should be folded neatly and ceremoniously.

Thinking about the US Flag, brings to mind various patriotic songs. Such as those found at this website:
https://www.scoutsongs.com/categories/patriotic-songs.html

The US Flag can be flown daily, but if flown at night it should be properly illuminated. The National Holidays that typically encourage flag flying are:
http://www.valleyforgeflag.com/43/flag-flying-holidays.htm

For a selection of **10 75 00** related LEED products, see:
http://www.arcat.com/divs/sec/sec107500.shtml

Flag Poles: ***Some additional information***

- Large American Flag Flies Again Over Penn College . . . *on a 120ft Flag Pole*
https://pctoday.pct.edu/large-american-flag-flies-again-over-penn-college/

10 81 00 - Pest Control Devices

Pest Control Devices includes Bird, Insect and Rodent control. Obviously, for health reasons and the protection from carpenter ants, termites, and any other destructive insects, it is important to have proper devices planned for and integrated into your project.

Not only is flashing important at the sill plate to prevent termites, but also at door and window heads, and all flashing that would prevent entry for insects and water. If water finds its way into a wood framed structure, or anywhere there might be wood, it is just a matter of time before decay and insect problems begin.

Pest Control Devices: *Some additional information*

- Identifying and Controlling Wood-Destroying Insects:
http://learningstore.uwex.edu/assets/pdfs/A3093.pdf

- Termite Control: Answers for Homeowners (Univ. of Kentucky)
https://entomology.ca.uky.edu/ef604

- Prevail Pest Control
http://www.prevailpest.com/termites.html

- Electronic Pest Control
http://www.pestproducts.com/electronicpestcontrol.htm

See the valuable below flashing details and descriptions from the Copper Development Association, Inc.

- Flashings and Copings: Through-Wall Flashing
http://www.copper.org/applications/architecture/arch_dhb/arch-details/flashings_copings/wall_flashing.html

- Flashings and Copings: Roof Penetrations
http://www.copper.org/applications/architecture/arch_dhb/arch-details/flashings_copings/roof_penetrations.html

- Flashings and Copings: Stepped and Chimney Flashings
http://www.copper.org/applications/architecture/arch_dhb/arch-details/flashings_copings/chimney.html

- Flashings and Copings: Counter Flashing
http://www.copper.org/applications/architecture/arch_dhb/arch-details/flashings_copings/counterflashing.html

- Flashings and Copings: Eave and Snow Flashing
http://www.copper.org/applications/architecture/arch_dhb/arch-details/flashings_copings/eave_flashing.html

- See ***Division 03, 04 and 07****, for related products and information.*

10 82 00 - Grills and Screens / Treillage

Solar Control Metal Mesh Screens allow for shading and Green wall Technology

- **Greenscreen®**
  1743 S. La Cienega Boulevard
  Los Angeles, CA 90035-4650
  Toll Free: 800.450.3494
  Phone: 310.837.0526

  http://greenscreen.com/products/elements/

  *"**greenscreen®** is a complete trellis system that includes a versatile array of engineered mounting hardware for almost every assembly."*

- **GKD Metal Fabrics**
  GKD-USA, Inc.
  825 Chesapeake Drive
  Cambridge, MD 21613
  Phone: 410.901.8433

  http://www.gkdmetalfabrics.com/blog/metal_mesh_cladding_keeps_the_suns_rays_at_bay.html

  "Façades are no longer just decorative for buildings; they are now providing substantial benefits in terms of sun control and LEED points. Woven metal mesh fabrics reduce solar-generated heat gain, while providing lighting benefits within a building. Solar management technology allows specifiers to plan ahead for the solar heat gain coefficient (the shading measurement used throughout the U.S.) of a building."

10 82 13 - Exterior Grills and Screens

Grills and Screens are necessary to ventilate, but should 'it' be ventilated? This debate as it relates to residential attic spaces is resolved below.

- **"To Vent or Not to Vent"**
  https://www.youtube.com/watch?v=Ld8pzIu45F8&nohtml5=False 53:09

  Thanks again to **Joe Lstiburek** at **Building Science Corporation**. He communicates the importance of a insulation, air-tightness and venting, of the ceiling and roof assembly.

- **PA-1101: A Crash Course in Roof Venting**
  http://buildingscience.com/documents/published-articles/pa-crash-course-in-roof-venting/view

- **Lamanco, Inc.**
  P.O. Box 519
  2101 West Main St.
  Jacksonville, AR 72076
  Toll Free: 800.643.5596

Phone: 501.982.6511
Fax: 501.982.1258

Lomanco West, Inc.
4450 Mohave Airport Road
Kingman, Arizona 86401

http://www.lomanco.com/

Since 1946, **Lamanco, Inc.** has been manufacturing quality ventilation products, including the *"often imitated, but never duplicated"* Vari-Pitch Gable Vent. See website for other various roof and foundation vents and an on-line ventilation calculator.

- ### Ventilation Products: Owens-Corning™ Roofing

  http://www.owenscorning.com/roofing/accessories/ventilation-products

End of Division 10

Space for notes:

DIVISION 11 - Equipment

This Division name is what it is, and to the point. The following sections offer various '***Equipment***' that are all very practical and important to equip our buildings with! Take a look.

11 12 00 - Parking Control Equipment

Parking Control Equipment obviously deals with vehicles. So, with the above, the below are some resources that communicate vehicle parking configurations as well as the stated requirement for the number of handicapped stalls.

- Parking Lot Design: Theory & Concepts (Univ. Idaho)
  http://www.webpages.uidaho.edu/niatt_labmanual/chapters/parkinglotdesign/theoryandconcepts/parkingstalllayoutconsiderations.htm

- Additional parking lot and site planning information: Chester Co. Planning Commission, (West Chester, PA) provides a variety of content for various applications including the code minimum number of ADA Accessible Parking Stall requirements.
  http://www.landscapes2.org/ToolsLandscape/Pages/ParkingDesign.cfm

Keep in mind, the local Zoning Department will need to review the Building Use proposal before anything else. Then, depending on the Building Use, and possible other requirements will determine the total overall parking stalls that the site should accommodate. If this number cannot be met, then a variance may be able to be approved through a Zoning Appeals process.

The Zoning Department may also have a landscape requirement as well, along with possible other items to be reviewed and approved. And, when proposing a new or revised Site Plan, do not neglect to accommodate the safety and accommodations of the pedestrian. The use of crosswalks (using a different material is nice, rather than simply painting crosswalk lines) and appropriate lighting is important. With the landscape potential requirement, curbed islands are the best way to simplify parking control in more of a permanent method rather than simply relying on all pavement striping. Not to mention this could be a way to provide bioswales for parking lot drainage. Then, in the regions where snow is common, don't forget about how and where snow is to be mounded, while not damaging the landscape. Details. Details. Details!

For a selection of **11 12 00** related LEED products, see:
http://www.arcat.com/divs/sec/sec111200.shtml

11 13 00 - Loading Dock Equipment

Since a majority of our goods in the USA are transported by a semi-truck , such a truck requirements need to be accommodated - be it loading docks themselves, as well as the trucks' turning radii. It is important to specify Dock Locks for trailers as Dock Safety is critical, especially with forklift transfers from dock to trailer.

- **Durable Corporation**
  75 N. Pleasant Street
  Norwalk, OH 44857
  Toll Free: 888.967.7628

  Phone: 419.668.8068
  Fax: 800.537.6287

  https://www.durablecorp.com/

  *"Since **Durable Corporation** was founded in 1923, our company has embraced a culture of recycling in our production process . . . for Dock Bumpers, Mats, and More."*

- **Rite-Hite**
  8900 N. ARbon Drive
  Milwaukee, WI 53223
  Toll Free: 888.214.0882
  Phone: 414.355.2600
  Fax: 414.355.9248

  www.ritehite.com

  *"After pioneering the vehicle restraint industry with the Dok-Lok in 1980, **Rite-Hite** has continually evolved trailer restraints to meet the industry's evolving challenges. This encompasses trailer and loading dock design along with regulatory compliance. The Rite-Hite Dok-Lok, also known as a truck restraint, trailer restraint, or dock lock, is known worldwide."*

Loading Dock Equipment: *Some additional information*

- **Loading Dock Planning Standards**
http://kelleyentrematic.com/assets/KelleyDockPlanningStandards.pdf

- **Loading Dock** (Whole Building Design Guide)
https://www.wbdg.org/design/loading_dock.php

- **Dok-Lok Rotating Hook Vehicle Restraints (RHR)**
https://www.youtube.com/watch?v=FBVrX92ZsmY 6:48

- **Loading and Unloading: OSHA**
https://www.osha.gov/SLTC/trucking_industry/loading_unloading.html

11 14 00 - Pedestrian Control Equipment

This section includes Portable Posts & Rails; Rotary Gates; Turnstiles; and Electronic Detection & Counting Systems.

For a selection of **11 14 00** related LEED products, see:
http://www.arcat.com/divs/sec/sec111400.shtml

11 23 00 - Commercial Laundry & Dry Cleaning Equipment

Per *BuildingGreen.com*, there has been increased *"energy efficiency of clothes washers in recent years . . . aided by the U.S. Department of Energy (DOE) minimum energy standards, the Federal Energy Star program, and efforts of the Consortium for Energy Efficiency (CEE)". "Clothes washers that use less* (hot) *water are usually more energy efficient, but . . . washing clothes in cold water is still the best energy-savings measure."*

For a selection of **11 23 00** related LEED products, see:
http://www.arcat.com/divs/sec/sec112300.shtml

11 24 13 - Floor and Wall Cleaning Equipment

There is nothing that is completely maintenance free. Cleaning and maintaining the buildings that we are entrusted is what is expected. When cleaning the floors and walls, it is best to select low-environmental-impact cleaning products, as well as natural ones that would ultimately contribute to the **Indoor Air Quality (IAQ)** and potentially reduce water pollution as well.

Antibacterial products are everywhere these days. They are in soaps, lotions, toys, even socks and underwear. While this may sound like a positive thing, many antibacterial and disinfectant products contain harmful chemicals that may be even more dangerous than the microorganisms they were meant to kill. It is understood that regular cleaning with plain soap and hot water will kill some bacteria. However, when "some" is not good enough, you can easily make your own natural disinfectants with just a few ingredients. For **natural cleaners and disinfectants,** this is just one of the many resources:
http://wellnessmama.com/6244/natural-cleaning-tips/

Along with above, the below is interesting . . .

- ✓ **White Vinegar** - excellent disinfectant. Kills 99% of bacteria, 82% mold, 80% viruses.
- ✓ White Vinegar with 3% **Hydrogen Peroxide** is even more effective.
- ✓ **Borax** - mix 2 tablespoons, 1/4 cup lemon juice and 2 cups hot water in a spray bottle for an all-purpose cleaner / disinfectant
- ✓ **Tea Tree Oil** (Health Food Stores) - a powerful antiseptic and fungus killer.

And the list goes on for healthier, less expensive options!

Triclosan, an Antibacterial product, can cause issues:
http://www.fda.gov/ForConsumers/ConsumerUpdates/ucm205999.htm

With vacuums, High Efficiency Particulate Air Filtration (**HEPA**) is best so not to spread the dust around in the air. Yet, add HEPA to a water separator process rather than a vacuum bag adds another unique step in filtration, as with the *Rainbow Cleaning System*.

- **Rainbow Cleaning System**

  Rexair World Headquarters
  50 West Big Beaver, Suite 350
  Troy , MI 48084
  Phone: 248.643.7222

  http://rainbowsystem.com/

*"The **Rainbow Cleaning System** cleans your home the most natural way possible, using The Power of Water®. It not only cleans your home... it creates a healthier living space."*

The **Rexair CSD** commercial version is capable to clean up toxic spills.
https://www.youtube.com/watch?v=84hqy2ocpl0 3:30

11 24 19 - Vacuum Cleaning Systems

As a 'system', this would include 'Central Vacuums'. Because I really do not have much experience with one, I do not have a preference. But, they appear efficient with the ease of simply connecting the hose to one of the various inlet ports and only needing to pull the hose assembly around. Meanwhile, the dust is collected in a remote centralized canister. It would seem that the inlet ports and internal piping would be subject to clogs from one of those stray socks or a child prank. Then what?!

I trust **Consumers Reports**. Per this link, it is unfortunate that they currently do not review Central Vacuums, but they have commented on other vacuum cleaners.

http://www.consumerreports.org/cro/vacuum-cleaners/buying-guide.htm

Is there truly a Central Vacuum made in the USA?

http://www.thinkvacuums.com/central-vacuum/made-in-usa.php

So, with the above website, the below manufacturer is hopefully honest with their claim . . .

- **Vacumaid**
  Lindsay Manufacturing, Inc.
  3 Darr Park
  Ponca City, OK 74601
  Phone: 800.546.3729

  http://vacumaid.com/

  "Made in the Ponca City, OK since 1956. Completely remove allergens and dust outside of your living environment. Easily installed in new or existing homes or businesses."

11 31 00 – Residential Appliances

Continuing on my trust for **Consumers Reports**, here is a link that relates to USA Made Appliances:
http://www.consumerreports.org/cro/magazine/2015/05/best-american-made-appliances/index.htm

And, thanks to *This Old House*, here is a related link discussing appliance life expectancy. *How Long Stuff Lasts.* Focusing on approximately how appliances will last—and approximately how much it will cost to replace them. See: **http://www.thisoldhouse.com/toh/article/0,,216991-4,00.html**

Once the appliance is no longer able to be repaired, or it would be less expensive in the long run to replace it rather than repair, see this recycle information: **https://www.energystar.gov/products/recycle**

I might add, that the refrigerator rebate program(s) offered by various utility companies seem a bit odd. Such a program encourages the consumer to turn over their old refrigerator for a *$35 Rebate* with the stipulation that the appliance is still in working condition. So, the consumer receives $35, but he/she now needs to purchase a replacement. Depending on size and its features, it could range from the mid $400's - to well over $1500, and more.

You might ask, with a new Energy Star refrigerator, how long will it take for the energy savings to pay for the refrigerator?

It is understood that this is encouraging the consumer to replace the working refrigerator with a new *Energy Star* appliance as a way to save on energy. But on the other side of the coin, it appears to be an economic stimulus effort (to make more refrigerators) while having consumers to perceive their appliance as being obsolete. Hmmm. Didn't we discuss previously **Perceived Obsolescence** and **Planned Obsolescence**?! Do a little research. See what you can find. Now, what happens to these old refrigerators? Are they recycled?

A few other examples of appliance Perceived Obsolescence includes the phasing out of the long favored Stainless Steel finish. A true test is if you have a new *Black* Stainless appliances, and even a new *Black* Stainless *Touch Screen* Refrigerator. Do you?!

When it comes to planning spaces for appliances, it is valuable to know their measurements. Some appliances vary in size, (like refrigerators), so check the specifications provided by the manufacturer.

Some appliances however are consistent in size, such as:

- 24" wide = Built-in Dishwasher (under counter)
- 30" wide = Slide-in Range (electric and natural gas)

For a selection of **11 31 00** related LEED products, see:
http://www.arcat.com/divs/sec/sec113100.shtml

*- See CSI **Division 12,** for related products and information for cabinetry - as appliances are being considered.*

11 30 33 - Retractable Stairs

Retractable Stairs, or otherwise known as Folding Stairs are a great concept and nice way to gain access to the attic space. Per the International Residential Code (IRC), **Chapter 8,** Section ***R807 - Attic Access***; and ***Chapter 11 - Energy Efficiency*** are important to refer to.

Use website to access the *I-Codes, PremiumACCESS or the Free PublicACCESS*: **https://codes.iccsafe.org/**

For a selection of **11 30 33** related LEED products, see:
http://www.arcat.com/divs/sec/sec113033.shtml

11 40 00 - Food Service Equipment

It is said that older Pre-Rinse Spray Valves at commercial kitchen sinks can use up to five gallons per minute (gpm), for up to four hours a day. Newly manufactured commercial Pre-Rinse Spray Valves can use as little as 1.15 gpm, or even .70 gpm, using higher pressure.
See: **http://www.fishnick.com/equipment/sprayvalves/**
(Food Service Technology Center)

For additional **11 40 00** related LEED products, see:
http://www.arcat.com/divs/sec/sec114000.shtml

11 51 00 - Library Equipment

- **Brodart**

  Brodart Library Supplies & Furnishings Division
  500 Arch Street
  Williamsport, PA 17701
  Toll Free: 800.265.8470
  Phone: 570.326.2461
  Fax: 800.283.6087

  http://www.brodart.com/

  *"Since 1939 libraries have been able to turn to **Brodart** for everything from shelf-ready books to electronic ordering systems, high quality furniture, and supplies. From humble beginnings, Brodart has grown to become an international company, serving libraries from the Northwest Territories to the Pacific Rim, with facilities in the United States and Canada. Brodart truly has evolved into a full-service library company."*

For additional **11 51 00** related LEED products, see:
http://www.arcat.com/divs/sec/sec115100.shtml

11 52 00 - Audio-Visual Equipment

- **Draper, Inc.**

  411 S. Pearl
  P.O. Box 425
  Spiceland, IN 47385-0425

  http://www.draperinc.com/default_us.aspx

  Draper manufactures Audio Visual & Presentation Products; Window Shades & Solar Control Solutions; and Gym & Athletic Equipment

For a additional **11 52 00** related LEED products, see:
http://www.arcat.com/divs/sec/sec115200.shtml

Audio-Visual Equipment: *Some additional information*

With much of our Audio-Visual Equipment made outside the United States, this website communicates those electronic items, including Home Theater equipment and other **American Made: Electronic Products or Services**.
http://www.madeinusa.org/nav.cgi?data/elec

11 53 00 - Laboratory Equipment

Laboratory Fume Hoods, as required by code. It is known that there is a significant amount of energy to not only exhaust the air, but then to condition and provide the make-up, replacement air for the space or building.

For a selection of **11 53 00** related LEED products, see:
http://www.arcat.com/divs/sec/sec115300.shtml

11 65 00 - Athletic and Recreational Equipment

- **Draper, Inc.**
  411 S. Pearl
  P.O. Box 425
  Spiceland, IN 47385-0425
  Toll Free: 800.238.7999
  Phone: 765.987.7999

  http://www.draperinc.com/default_us.aspx

  Draper manufactures Audio Visual & Presentation Products; Window Shades & Solar Control Solutions; and Gym & Athletic Equipment. Draper has been manufacturing since 1902.

For a additional **11 65 00** related LEED products, see:
http://www.arcat.com/divs/sec/sec116500.shtml

11 68 00 - Playfield Equipment & Structures

For Playground Structures, products should be made from recycled plastic, recycled metal and/or sustainably harvested wood.

- **Cedarworks**
  799 Commercial Street
  PO Box 990
  Rockport, ME 04856-0990
  Toll Free: 800.462.3327
  Phone: 207.596.1010

  https://www.cedarworks.com/

  "At CedarWorks our commitment is to the highest quality construction. We use time tested construction methods and an

unwavering attention to detail to make sure that every CedarWorks, indoor or outdoor, is built to last."

- **Playworld, Inc.**
  1000 Buffalo Road
  Lewisburg, PA 17837-9795
  Toll Free: 800.233.8404
  Phone: 570.522.9800

  http://playworld.com/

  "Serving parks and recreation departments, municipalities, schools, early childhood daycare facilities, churches and others; ***Playworld*** *continues our tradition of creating meaningful play for today's society."*

- **Rainbow Play Systems, Inc.**
  Phone: 800.RAINBOW

  http://www.rainbowplay.com/

  Multiple Showroom locations.

  *"**Rainbow** is the industry leader having the largest playground factory in the U.S.A and over three decades as an American corporation. Rainbow has over 200 locations worldwide to serve you, providing professional delivery & installation service nationwide and offers the best Lifetime Warranty coverage in America."*

For additional **11 68 00** related LEED products, see:
http://www.arcat.com/divs/sec/sec116800.shtml

11 70 00 - Healthcare Equipment

For a selection of **11 70 00** related LEED products, see:
http://www.arcat.com/divs/sec/sec117000.shtml

- *See CSI Section **10 25 00 - Service Walls,** for related products and information.*

And, in a related topic, a study at **Johns Hopkins Hospital** revealed that hands-free faucets harbor germs - which ties back to the bacteria topic mentioned in *CSI Section **11 24 13***.
http://articles.latimes.com/2011/mar/31/news/la-heb-hands-free-faucet-bacteria-20110331

*See CSI Section **22 40 00 - Plumbing Fixtures,** for related products and information.*

11 82 00 - Recycling and Solid Waste Handling Equipment

To be good environmental stewards, enabling and encouraging easily understood resource recycling within buildings is a must. From the recycling station to being bailed at the Recycling Center is a **LEED Materials and Resources Prerequisite** expectation. **http://www.usgbc.org/credits/new-construction/v4/material-%26-resources**

Lycoming County Waste and Resource Management - (Pennsylvania)

http://www.lyco.org/Departments/ResourceManagementServices.aspx
with information video(s)

It is stated that the cogeneration created land fill gas, converted to electricity, is sold to the local utility company.

For additional **11 82 00** related LEED products, see:
http://www.arcat.com/divs/sec/sec118200.shtml

End of Division 11

Space for notes:

DIVISION 12 - Furnishings

As we know, ***Furnishings*** 'make the space'. They include the actual room furniture along with all the other necessary and potential memorable items found in the following sections that make the room, a home, an office, the classroom, a doctor's office or any space more than simply functional.

This Proverb comes to mind . . .

> *By wisdom a house is built,*
> *And by understanding it is established;*
> *And by knowledge the rooms are filled*
> *With all precious and pleasant riches.* Proverbs 24:3-5

It is unfortunate though that some of these items are discarded shortly after purchase because they are made cheaply and wear out, break or does not work after a time, or ???. This again reminds me of the **Planned Obsolescence** and **Perceived Obsolescence** topic discussed within the ***Introduction*** of this book, and elsewhere. And, along with this topic, some of the furniture that is made from particle board (press board, or some other name), can commonly contain formaldehyde in the glue. If you were not aware, formaldehyde is one of those chemicals that cause poor indoor air quality (IAQ) and can contribute to **Sick Building Syndrome.** It is very unfortunate, but if you were not aware, the use of formaldehyde resulted in poor indoor air quality, which then resulted in various health problems when the **Federal Emergency Management Agency (FEMA)** supplied temporary trailer housing to the 2005 Hurricane Katrina victims. This ultimately led to a $42.6 million dollar class-action law suit that affected approximately 55,000 people! Just terrible. See:

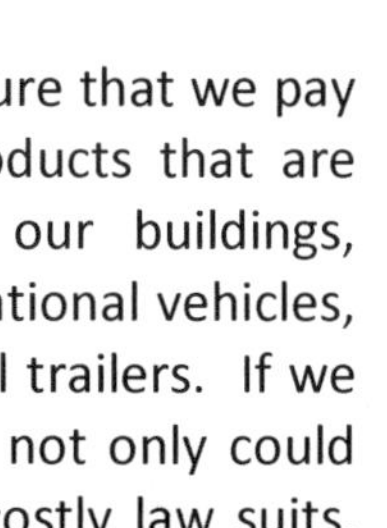

http://www.cbsnews.com/news/katrina-rita-victims-get-426m-in-toxic-fema-trailer-suit/

This is a reminder to all of us to make sure that we pay attention to all the materials and products that are specified and used to manufacture our buildings, including temporary housing and recreational vehicles, as innocently used in these recreational trailers. If we do not restrict such noxious products, not only could future health issues arise, along with costly law suits, but such negligence would cause such tainted manufactured goods to be disposed of in our landfills! Natural disasters, such as flooding, tornadoes and hurricanes cause enough debris to be transported to these landfill sites, so it would be best that such products were either **Cradle to Grave** (C2G), without contaminants, or even better, **Cradle to Cradle** (C2C). This would permit our debris to be recycled back to nature or the marketplace.

Before we continue, (and as a continuation of CSI Section 11 82 00 - Recycling and Solid Waste Handling Equipment), take a moment to see where the largest landfills are in the US. See:
http://www.cnbc.com/2010/09/27/Americas-Largest-Landfills.html

Here is a related landfill link from the **US Environmental Protection Agency** (EPA):
https://www.epa.gov/landfills

12 10 00 - Art

This section includes paintings, murals, sculpture, and other creative works that adorn our buildings. See the following sections for related information . . .

12 11 00 - Painted and Photo Murals

Such painted and photo based murals are applicable for interior walls, floors and ceilings and exterior applications.

A perfect example of such murals would be what is known as ***Trompe-l'œil.*** A *French* term that means to 'fool the eye', and is pronounced *'Trum-ploy'*. A few of the many artists include:

- **Art Effects**
  Email: twohofmanns@gmail.com
  Phone: 863.521.3356
  http://www.trompe-l-oeil-art.com/trompe.html

- **Doug Rouse**
  http://www.rouse66.com/about/

*- See CSI Section **09 90 00 – Painting and Coating,** for related products and information.*

12 14 00 - Sculptures

- **National Sculptures' Guild**
  2683 N Taft Avenue
  Loveland, Colorado 80538
  Phone: 970.667.2015

  http://www.nationalsculptorsguild.com/

12 17 00 - Art Glass

*- See CSI Section **08 81 13 – Decorative Glass Glazing,** for related products and information.*

12 19 00 - Religious Art

- **Religious Art - The Definitive Collection**
  http://www.religious-art.org/

For a selection of **12 19 00** related LEED products, see:
http://www.arcat.com/divs/sec/sec121900.shtml

12 20 00 - Window Treatments

Window treatments are the window coverings on the interior side of a window. They can include curtains, drapes, valances, blinds, screen mesh or opaque shades, interior shutters and window quilts. Such items provide privacy, solar shading, and thermal solutions. These are typically included with the sale of an existing home or business.

- **Draper, Inc.**
  Corporate Headquarters
  411 South Pearl Street
  P.O. Box 425
  Spiceland, IN 47385
  Toll Free: 800.238.7999
  Phone: 765.987.7999

  http://www.draperinc.com/default_us.aspx

  Draper manufactures Audio Visual & Presentation Products; Window Shades & Solar Control Solutions; and Gym & Athletic Equipment. Draper has been manufacturing since 1902.

- **MechoSystems**
  42-03 35th Street
  Long Island City, NY 11101
  Phone: 718.729.2020
  Fax: 718.729.2941

  http://mechoshade.com/

  *"**MechoSystems** has been the leading innovator of manual, motorized, and automated solar-shading and room-darkening solutions for more than 50 years."*

  The MechSystems **EcoVeil® Screens** are Cradle to Cradle and GreenGuard Certified.

- **Window Quilt Insulated Shades**
  22 Browne Court, #105
  Brattleboro, VT 05301
  Phone: 802.246.4500

  https://www.windowquilt.com/products/full_broch.htm

  *"(With) over 600,000 commercial and residential installations . . . **Window Quilt** performs equally well in summer and winter by countering all four heat transfer mechanisms. Think of them as moveable insulation for windows."*

For additional **12 20 00** related LEED products, see:

http://www.arcat.com/divs/sec/sec122000.shtml

If you are looking for other alternatives to window treatments, you might consider *CSI Section **08 88 36 - Switchable (Smart) Glass***, for privacy reasons, and/or *CSI Section **08 87 13 - Window Film** (Glazing Film).*

Also, if there is a need to remove a window, many times the first thought is to try to patch the exterior finish. However, what is done in many places is to use exterior window shutters to cover the window opening. Then, on the interior side, the wall can be finished as necessary. Historically this concept was even done as an original design concept, to maintain a symmetry to the facade, with the intentions to never install the window.

12 30 00 - Casework

Casework refers to the cabinetry and other custom built-in furniture that can be found in residential and non-residential buildings. A few examples could include the front Reception desk, nurses stations, break room cabinets, science laboratory cabinetry, kitchen cabinets, exterior stainless steel cabinets, bookcases and display cabinetry, including a full or partial wall entertainment center with built-in wall shelving that could include a 'Murphy wall-bed' and a home office work space. Such custom design opportunities would create a space that is flexible to many uses. You might agree that such flex-space ideas are important to good design. What can you design or make?!

Many times, to minimize the costs (including pollution) related to transportation, and to promote local business, it is preferred to find and use locally milled wood, along with a local cabinet manufacturing company, or an individual craftsman that is well known in the area.

Casework typically is built to standard sizes. For base cabinetry the widths are commonly in 3" increments. The height is commonly 34.5" without the counter added. The depth of the cabinet is commonly is 24". For wall cabinets, the widths are also commonly 3" increments to match the base cabinets. The heights vary in 3" increments as well, with 30" tall cabinets as being most common, placed 18" above the countertop. The depth of the cabinet is commonly is 12"

It is important to understand that the above heights are typically adjusted to accommodate **Universal Design** expectations, understanding every person's needs and abilities. If in a wheelchair for instance, the counter height is typically 2" less, or otherwise 34" above the finished floor.

Then, if not understood, it should be stated that all cabinetry can be custom made to meet the project requirements. Sometimes this can add cost to the project, but in many cases, the quality is better and the price can be less. This is where the designer or building owner needs to 'shop around' for sustainably made casework.

See the below resources:

- **Cabinet Makers Association**
  47 W. Polk Street
  Suite 100-145
  Chicago, IL 60605-2085
  Phone: 866.562.2512
  Fax: 866.645.0468

  http://www.cabinetmakers.org/

  "The Cabinet Makers Association is a professional organization where cabinetmakers and woodworkers from both the

residential and commercial markets get together and share their hard earned knowledge and experience to help one another."

- **Kitchen Cabinet Manufactures Assoc.** (KCMA)
  1899 Preston White Drive
  Reston, VA 20191-5435
  Phone: 703.264.1690
  Fax: 703.620.6530

  http://www.kcma.org/

  *"The **Kitchen Cabinet Manufacturers Association (KCMA)** is a voluntary, non-profit trade association representing North American cabinet manufacturers and suppliers to the industry. KCMA is an influential advocate for the industry and since 1955 has administered the nationally-recognized performance standard for cabinets (ANSI/KCMA A161.1). Today, KCMA also is leading the way in promoting responsible environment practices in the industry.*

  Whether you are a manufacturer, designer, architect or homeowner, this site is your go-to resource for all types of cabinetry information."

- **National Kitchen & Bath Association** (NKBA)
  687 Willow Grover Street
  Hackettstown, NJ 07840
  Phone: 800.843.6522

  https://www.nkba.org/Design.aspx

  The National Kitchen & Bath Assoc. (NKBA) provides training, certification along with a network of Certified design professionals and other resources for the building/home owner and professional.

For a selection of **12 30 00** related LEED products, see:
http://www.arcat.com/divs/sec/sec123000.shtml

12 36 00 - Countertops

If base cabinetry is at a desk or various counter work heights, then of course there is the need for a countertop surface. Counter materials range in cost, appearance, durability and maintenance.

The various counter surfaces options typically include: Concrete, Laminate, Quartz, Solid Surface (Paper Composite), Solid Surface (Plastic), Stainless Steel, Stone, Terrazzo (Recycled Glass), Ceramic Tile, Glass Tile, and Wood. See the following CSI Sections and the below website for a description of these options:
Choose the Best Countertop Material for Your Home and the Environment - **http://www.greenhomeguide.com/know-**

how/article/choose-the-best-countertop-material-for-your-homeand-the-environment#laminates

And, as discussed previously, be careful to not be trendy. After the expense of installing a new countertop, recognize that even countertops can be trendy. Then what? Even if the product selected was the most sustainable product today, it can be considered no longer 'in-vogue', so the pressure would be to replace it in a few years. As you know, this would be an example of **Perceived Obsolescence**. See the following link: *Good Bye to Granite countertops: The 6 Hottest Countertop Finishes* **http://dishwashers.reviewed.com/features/goodbye-granite-the-6-hottest-countertop-finishes?utm_source=TB_Paid&utm_medium=cpc&utm_campaign=tribunedigital-chicagotribune**

12 36 13 - Concrete (counters):

- **Portland Concrete Association** (PCA)
  5420 Old Orchard Road
  Skokie, IL 60077-1083
  Phone: 847.966.6200
  Fax: 847.966.9781

  1150 Connecticut Ave, NW, Ste. 500
  Washington, DC 20036-4104
  Phone: 202.408.9494
  Fax: 202.408.0877

  http://www.cement.org/for-concrete-books-learning/materials-applications/architectural-and-decorative-concrete/concrete-countertops

- **The Concrete Countertop Institute**
  P.O. Box 25898
  Raleigh, NC 27607
  Toll Free: 888.386.7711
  Phone: 919.275.2121
  Fax: 919.882.9700

  http://www.concretecountertopinstitute.com/

  *The **Concrete Countertop Institute's** mission is to raise the standard for concrete countertops. Contact the institute for training, product information, and images of beautiful do-it-yourself artistic concrete projects.*

For a selection of **12 36 13** related LEED products, see:
http://www.arcat.com/divs/sec/sec123613.shtml

12 36 23 - Plastic Laminate (counters):

- **Formica**

Formica Group
1379 North Delaware Drive
Mt. Bethel, PA 18343
Phone: 570.897.6319

https://www.formica.com/en/us

*"**Formica** Group globally leads the industry in the design, manufacture and distribution of innovative surfacing products for commercial and residential applications. Formica Group is committed to making sustainable principles and practices a part of everything we do. We strive to adhere to the highest ethical standards as we advance in our efforts to protect vital resources for future needs."*

- ## Nevamar

Panolam Surface Systems
20 Progress Drive
Shelton, CT 06484
Phone: 203.925.1556

http://www.nevamar.com/

*"**NEVAMAR®** HPL is the brand name for our line of durable, high performance, high pressure laminates (HPL). NEVAMAR® HPL continues its long-standing tradition of design and technical innovation featuring a broad selection of quality products. Today, our High Pressure Laminates are more versatile than ever. Simply look at our selection of over 20 specialty laminates and you'll know exactly what we mean."*

- ## Wilsonart

Wilsonart Engineered Surfaces
2501 Wilsonart Drive
P.O. Box 6110
Temple, TX 76503-6110
Phone: 800.433.3222

http://www.wilsonart.com/

*"**Wilsonart** Engineered Surfaces is a world leading manufacturing and distribution organization that is driven by a mission to create innovative, high performance surfaces people love, with world-class service our customers can count on, delivered by people all over the world who care.*

Our company, headquartered in Austin, Texas, manufactures and distributes High Pressure Laminate (HPL), Quartz, Solid Surface, Coordinated TFL and Edgebanding and other engineered surface options for use in furniture, office and retail spaces, countertops, worktops and other applications."

For additional **12 36 23** related LEED products, see:
http://www.arcat.com/divs/sec/sec123623.shtml

12 36 40 - Stone Countertops

Stone Countertops typically include Marble, Granite, Slate and Soapstone. Each have their place and preference, but recognize that Soapstone is the only one listed that is not porous, so it is naturally stain resistant per various sources, including the American Institute of Architects. See what else is stated . . .

http://www.aia.org/akr/Resources/Documents/AIAP037034

- ## The Natural Stone Institute

Ohio Office
380 E. Lorain Street
Oberlin, OH 44074

New York Office
36 Cougar Lane
P.O. Box 419
Chestertown, NY 12817

Phone: 440.250.9222
Fax: 440.774.9222

https://www.marble-institute.com/

"MIA+BSI: the ***Natural Stone Institute*** *serves more than 1900 members in 55 countries who represent every aspect of the natural stone industry, offering them a wide array of technical and training resources, professional development, regulatory advocacy, and networking events. Two prominent publications—the* DIMENSION STONE DESIGN MANUAL *and* BUILDING STONE MAGAZINE—*raise awareness in both the industry and the design communities for the promotion and best use of natural stone."*

MIA (Mable Institute of America) + **BSI** (Building Stone Institute)

- ## Old House Web

Soapstone suppliers
http://www.oldhouseweb.com/suppliers/ceramic-tile-and-stone/soapstone/

- ## Vermont Soapstone Company

248 Stoughton Pond Road
Perkinsville, VT 05151-0268
Phone: 866.572.9232
Fax: 802.263.9451

http://vermontsoapstone.com/visit-our-showroom/

"Since 1856, the craftsmen at ***Vermont Soapstone*** *have been handcrafting custom fixtures and home accents in our Perkinsville, VT facility. Originally producing primarily bed warmers, boot dryers, heat registers, griddles and wood burning stoves, today Vermont Soapstone is widely known for our flooring, countertops and wide, deep sinks."*

For additional **12 36 40** related LEED products, see:
http://www.arcat.com/divs/sec/sec123640.shtml

Stone Countertops: *Some additional information*

- ## Alkemi

Renewed Materials

P.O. Box 55
Cabin John, MD 20818
Phone: 877.462.6020
Fax: 301.320.3341

http://www.designguide.com/products/67027/Alkemi-by-Renewed-Materials

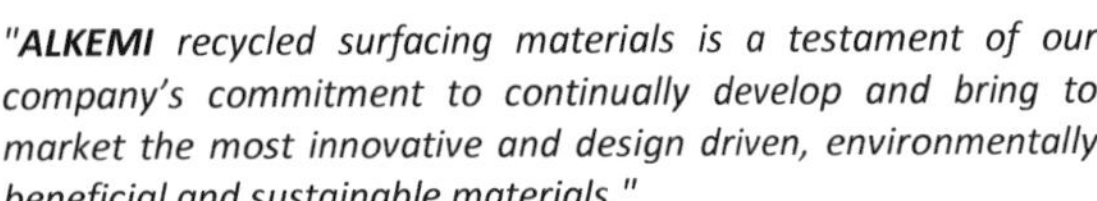

*"**ALKEMI** recycled surfacing materials is a testament of our company's commitment to continually develop and bring to market the most innovative and design driven, environmentally beneficial and sustainable materials."*

- **EnviroSLAB, Inc.**
  1107 New Hope Road
  Raleigh, NC 27610
  Toll Free: 800.849.8849
  Phone: 919.386.6864

  http://www.enviroslab.com/

  *"**EnviroSLAB Inc.** specializes in the manufacturing of terrazzo slab products utilizing 100% recycled content. Affordable and unique **EnviroSLAB** countertop slabs are custom manufactured with 100% recycled glass and/or porcelain and color-customizable epoxy resin in Raleigh, NC."*

12 40 00 - Furnishings and Accessories

This section includes Office and Desk Accessories; Bath Furnishings (including the bath mats, towels, and shower curtains; Bedroom Furnishings; along with various decorative and useful accessory items found in a space.

For a selection of **12 40 00** related LEED products, see:
http://www.arcat.com/divs/sec/sec124000.shtml

12 48 00 Rugs and Mats

In contrast to carpeting, as discussed in Division 09 - Finishes, this section addresses what sometimes are known as 'area rugs' and floor mats. As mentioned specifically in CSI Section **09 68 00**, carpeting can contribute to poor Indoor Air Quality (IAQ), as it traps dirt. Other than actually taking off your shoes, (as done in some cultures and personal choice) an **area rug** or a **walk off entrance mat** placed at the buildings entry door(s) are essential as a way to wipe off shoes and leave the dirt at the door.

And, as we know, **floor mats** are also used as a way to relieve foot, leg and back fatigue if placed at a location where someone will stand for a period of time.

Along with the above, **carpet runners** commonly used for stairs and corridors are also included in this section.

- **American Floor Mats**

152 Rollins Ave., #102
Rockville, MD 20852
Toll Free: 800.762.9010
Fax: 240.780.3309

https://www.americanfloormats.com/green-mats-recycled-mats/

*"**American Floor Mats** brings you 25 years of floor mat experience. We are proud to offer you the most comprehensive selection of high quality floor mats such as: custom floor mats and logo floor mats; entrance mats, door mats, and welcome mats; anti fatigue mats, anti static mats, rubber mats, gym mats, chair mats, kitchen mats, pool mats, sticky mats, safety mats, and bathroom mats."*

- **Consolidated Plastics**

4700 Prosper Drive
Stow, OH 44224
Toll Free: 800.362.1000

https://www.consolidatedplastics.com/Commercial-Matting.aspx

"Consolidated Plastic has a large selection of various Commercial Mats. We carry a wide selection of commercial matting, everything from standard entry way floor mats to Custom Logo Mats. In addition, Consolidated Plastics offers many anti-fatigue mats for worker comfort and safety."

- **Construction Specialties (C/S)**

C/S Muncy, PA
6696 State Route 405
Muncy, PA 17756
Toll Free: 800.233.8493
Phone: 570.546.5941

http://www.c-sgroup.com/entrance-flooring/entrance-mats

Construction Specialties has multiple locations and each manufacturing facility specializes in a specific product line. The Muncy, PA location manufactures and distributes: *"Acrovyn, Entrance Flooring & Expansion Joint Covers"*

- **The Carpet and Rug Institute (CRI)**

Street Address:
100 South Hamilton Street
Dalton, GA 30720

Mailing Address:
P.O. Box 2048
Dalton, GA 30722-2048

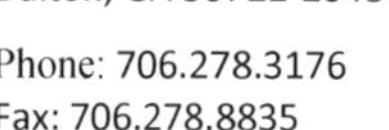

Phone: 706.278.3176
Fax: 706.278.8835

http://www.carpet-rug.org/

*"Today, indoor air quality (IAQ) is an important environmental consideration, especially since we spend approximately 90 percent of our time indoors. In 1992, **CRI** launched its Green Label program to test carpet, cushions and adhesives to help specifiers identify products with very low emissions of Volatile Organic Compounds (VOCs)."*

12 48 50 – Fitness Room Flooring & Mats

Typically Fitness Room integrate a rubber or synthetic rubber interlocking tile floor system.

As with any building, It is important to review potential code requirements. See the International Building Code (IBC) website for **Chapter 8** - *Interior Finishes.*

Use website to access the *I-Codes, PremiumACCESS or the Free PublicACCESS*: **https://codes.iccsafe.org/**

For a selection of **12 48 50** related LEED products, see:

http://www.4specs.com/s/12/12-4850.html

*See also CSI Section **09 65 40 – Rubber Flooring** and **09 68 00 – Carpeting,** for related products and information.*

12 50 00 - Furniture

While designing residential spaces, it is important to understand the *Minimum Room Areas* permitted and discussed in the International Residential Code (IRC). See IRC Section **R304**.

For non-residential spaces, it is important to reference the International Building Code (IBC) (Chapter 10) **Table 1004.1.1** (2009) or **Table 1004.1.2** (2012, 2015), to learn the *Maximum Floor Area Allowances Per Occupant*. This becomes a fantastic design tool as you might wonder for instance, what would be the minimum size that an office would need to be for four employees? This is just one example of how important the building code becomes your friend and not the enemy.

Along with the above, it is also important to consider furniture sizes in respect to the room dimensions, along with people circulation, door and window placement (as it relates to wall space), door widths (for emergency egress, accessibility and circulation paths), along with minimum window sizes for egress, daylighting, ventilation, not to mention the various potential views seen from different seating heights. See the International Building Code (IBC) website:

- **Chapter 10** - *Egress*
- **Chapter 11** - *Accessibility*
- **Chapter 12** - *Interior Environment*

See International Residential Code (IRC), **Chapter 3**:

- Section ***R303 - Light, Ventilation & Heating***
- Section ***R310 - Emergency Escape & Rescue Openings***
- Section ***R311 - Means of Egress***

. . . And potentially others.

Use website to access the *I-Codes, PremiumACCESS or the Free PublicACCESS*: **https://codes.iccsafe.org/**

See the beginning of ***Division 14 - Conveying Equipment*** *for Universal Design related information.*

12 59 00 - Systems Furniture

Systems Furniture represents the various USA manufactures of quality ergonomically designed modular office furniture that include Panel-Hung Component Furniture and Free-Standing Component Furniture.

- **American Furniture Systems**
  808 W. Santa Anita Ave.
  San Gabriel, CA 91776
  Phone: 626.457.9995

http://www.americanfurnituresystems.com/

"We manufacture all furniture items right here in the United States using only the best American-based raw materials."

*"**American Furniture Systems** is proudly known as a company that conforms to the highest standards of manufacturing. We built our reputation throughout the design and dealer community by proving that we stand behind every product we make, and treat every project, regardless of the size, with equal consideration. Our job isn't just to manufacture the best corporate furnishings on the market. It's to ensure that your*

clients are pleased with the products you recommend and to help you win their trust. We can only do that by winning yours."

- **Herman Miller, Inc.**

855 East Main Ave.
P.O. Box 302
Zeeland, MI 49464-0302
Phone: 616.654.3000

http://www.hermanmiller.com/products/workspaces.html

"We operate through several focused businesses, brands, and distribution channels, including Herman Miller, Herman Miller Healthcare, Nemschoff, Geiger International, and independently owned dealerships. All of them work to design and build a better world around you."

- **Knoll**

1235 Water Street
East Greenville, PA 18041
Toll Free: 800.343.5665

http://www.knoll.com/design-plan/products/by-category/systems-furniture

Knoll designs and manufactures furniture for Healthcare; Small Business and Startups; Higher Education; Government/GSA; Canadian Federal Government; Legal; Global Business Division; and Residential

- **Steelcase**

Global Headquarters
901 44th Street SE
Grand Rapids, MI 49508-7594
Phone: 616-247-2710

https://www.steelcase.com/

"We believe that providing the best solutions for our customers begins by ensuring they're the best solutions for our environment. That's why every step of the way – through design, manufacturing, delivery and product lifecycle – we consider the impact of our work on people and on the environment and uncover opportunities to make things better."

12 60 00 - Multiple Seating

This section includes: Fixed Audience Seating; Portable Audience Seating (such as folding chairs & interlocking stacking chairs); Stadium and Arena Seating, Telescoping Stands and Bleachers; along with Pews and Benches.

With multiple seating, a few code references comes to mind. Use the above link to access IBC:

- Section **303** for how seating helps to define various Assembly Groups for building use.

- Section **1004** for how seating affects Occupant Load (otherwise understood as how many occupants can fit safely in a space or building).
- Section **1004.4** – to learn the minimum seating length designated for fixed seating without arms, and seating booths with back rests.

For a selection of **12 60 00** related LEED products, see:
http://www.arcat.com/divs/sec/sec126000.shtml

12 90 00 – Other Furnishings

- **Americans Working.com**

  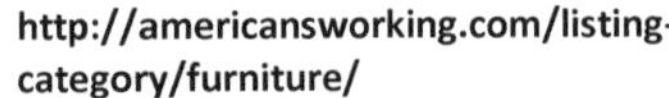

  http://americansworking.com/listing-category/furniture/

  Americans Working.com provides an updated, on-going list of American Made furniture companies.

- **Flexsteel**

  107 Pitney Road
  Lancaster, PA 17602
  Phone: 717.392.4161

  http://www.flexsteel.com/

  "American-Made Furniture for Home and Business"

  All of Flexsteel's stakeholders have a responsibility to protect our employees and our environment. The officers of Flexsteel and its subsidiaries will use our role as business and community leaders to set the tone at the top to guide our management teams in their efforts to improve the workplace and the environment we directly impact.

- **Reinbarnation**

  377 River Bend Road
  Siler City, NC 27433
  Phone: 919.542.4937

  http://www.reinbarnation.com/index.html

  "Furniture, Frames, Home & Garden Accessories created from reclaimed North Carolina barn wood."

  *"When old barns reach the end of their functional lives they are often demolished, buried, or burned to make way for new land uses. Roger Dinger, through **Reinbarnation,** offers a more positive destiny for these picturesque structures by reclaiming, repurposing, and reusing the vintage wood in his art.*

  The past life of each barn is honored prior to deconstruction. The building and grounds are explored, photographs are taken, and the owner's knowledge of the barn's past is recorded."

- **Sustainable Composites, LLC.**

1866 Colonial Village Lane, Ste. 109
P.O. Box 10371
Lancaster, PA 17605
Phone: 717.945.5791

http://www.sustainablecomp.com/

*"**Sustainable Composites** provides manufacturers and retailers with environmentally friendly materials produced using environmentally concerned processing. Our scientists have discovered a new, environmentally conscious, sustainable composite we call enspire leather. Using waste from tanneries that would normally go to landfills, enspire, is an environmentally and economically smart alternative to traditional leather. enspire contains 100% pre-consumer recycled leather fiber, which has the potential to be used for any application in which leather is typically used. And delivers the exact same properties!"*

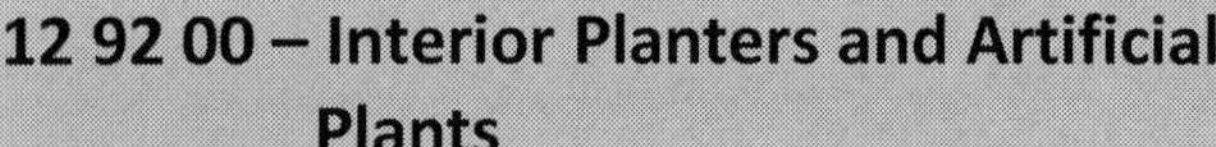

12 92 00 – Interior Planters and Artificial Plants

Interior Planters and Artificial Plants sounds nice, but in this day and age of awareness of **Indoor Air Quality (IAQ)**, why artificial plants? We all know that *real* live plants provide air purification qualities and other benefits. So, it is up to the architectural designer to provide the necessary daylighting (per code for its occupants), but also allows the live plants to grow and the building personnel to maintain them.

How it's Made - Artificial Flowers
https://www.youtube.com/watch?v=Vj1Id7VXk-Q 4:59
The above video shows the making of some beautiful artificial orchids. *But, how sustainable is this?*

Now, compare artificial plants with what this website says about all the benefits of real live plants indoors, and elsewhere.
https://greenplantsforgreenbuildings.org/

For a selection of **12 92 00** related LEED products, see:
http://www.arcat.com/divs/sec/sec129200.shtml

12 93 00 – Site Furnishings

This section includes Bicycle Racks (which of course encourages the benefits of exercise and an alternative form of transportation); Trash and Litter Receptors (but, should also be specific and include recycling containers); Manufactured Planters (for the exterior); plus, Site Seating and Tables.

- **Premier Polysteel**
  305 Enterprise Drive
  Northwood, IA 50459
  Toll Free: 877.746.3826
  Phone: 641.324.2750
  Fax: 641.324.2750

  http://www.premierpolysteel.com/

  Premier Polysteel manufactures commercial grade exterior Tables; Kids Tables; Benches, Receptacles; Bike Racks; Patio Furniture; Swings; Umbrellas; Grills; Planters; and Tree Grates & Guards

- **Telescope Casual Furniture**
  82 Church Street
  Granville, NY 12832
  Toll Free: 800.642.4645
  Fax: 800.458.3876

  http://www.telescopecasual.com/

  *"**Telescope Casual** has been producing quality outdoor patio furniture for over a century. Our product line includes wicker, aluminium, cast aluminium and resin furniture. Our products feature the highest quality of any porch and patio furniture or accessories in the industry."*

For additional **12 93 00** related LEED products, see:

http://www.arcat.com/divs/sec/sec129300.shtml

End of Division 12

Space for notes:

DIVISION 13 - Special Construction

As noted in the following sections, this **Special Construction** Division offers a variety of unique or special features and opportunities, including Swimming Pools; Fountains and waterfalls (for interior and exterior); Aquariums; Amusement Park Structures and Equipment; Tubs and Pools; Ice Rinks; Kennels & Animal Shelters; and the list goes on as seen in the following sections.

On the next page, there is a bulleted list of **Alternative Materials** that are *special* or unique. Some of these are already covered in specific CSI Sections, but there are those that are not. Either way, as a way to at least group these together, this Division seems to be the most appropriate place to do so, even though this really is not technically correct.

Alternative Construction / Alternative Materials -

Alternative Construction typically is possible by using Alternative Materials. In several instances alternative materials that may appear 'more green' than a traditional materials are not commonly accepted because they are 'new' or otherwise non-traditional, at least currently in our Country. With this they do not have the 'track-record' or the familiarity that others do. Many of the traditional materials have gone though expensive testing, evaluation, research and even political bargaining to become what we know today. And, it is a result of learning from past disasters that the Building Code have been developed, refined and expanded upon to the point that it offers very prescriptive measures, referencing such materials or assemblies, (especially those that do not combust or that have a high compressive or tensile qualities for the health, safety and welfare of the occupants), to ultimately prevent past problems or disasters from happening again!

However, the Building Code does provide a section that gives 'wiggle-room' for such alternative materials to be considered, and possibly approved by the Authority Having Jurisdiction (AHJ) for use - as long as the said alternative product or system proposed is *". . . satisfactory and complies with the intent of the provisions of this code, and that the material, method or work offered is, for the purpose intended, at least the equivalent of that prescribed in this code in quality, strength, effectiveness, fire resistance, durability and safety."* (Per IBC Section **104.11** and similar to IRC Section **R104.11**)

Use website to access the *I-Codes, PremiumACCESS or the Free PublicACCESS*: **https://codes.iccsafe.org/**

Alternative Materials that allow for **Alternative Construction** include:

- **Adobe** - CSI **Section 04 24 00**
- **Cob**
- **Rammed Earth Walls** - CSI **Section 13 35 13**
- **Poured Earth**
- **Earthbag** – "super adobe"
- **Strawbale** - Load Bearing and non-load bearing
- **Cordwood Masonry Construction**
- **Timber Frame** - CSI **Section 06 13 23**
- **Bamboo**
- **Earthship**
- **Papercrete**
- **Lightweight Concrete**
- **Stone** - CSI **Division 04**
- **Hybrids** (combining these in some way)
- **Manufactured Systems** - (includes *Autoclaved Aerated Concrete (ACC)* - CSI Section **03 41 00**)

AAC/ACC: *Some additional information*
Autoclaved Aerated Concrete *(AAC),* is sometimes known as ***Autoclaved Cellular Concrete*** *(ACC)* and even ***Aircrete***. *See CSI Section **03 41 00** for related information.*

One of the various websites that addresses many of the above Alternative Materials is: **http://www.greenhomebuilding.com/**

Along with the above, the ***Papercrete*** manufacturing and promotion by Barry J. Fuller along with the natural beauty of *Rammed Earth* are worth additional information.

- **Papercrete** - (Barry J. Fuller, Tempe, AZ)
  At the below website, Mr. Fuller communicates various projects and Papercrete courses that he offers - to share his experience in recycling newsprint, junk mail and magazines to make an

alternative building material. **http://www.livinginpaper.com/**

The below video communicates how Mr. Fuller uses Papercrete to build *Paper **Palace One***, a 500 SF office.
https://www.youtube.com/watch?v=eM34F4ZSmXQ 2:55

Papercrete Flame Test -
https://www.youtube.com/watch?v=lVipiNDaBYU 3:01

- **Rammed Earth** -
  Rammed Earth uses a combination of cement and Earth while the mixture is compressed (or rammed) into layers (lifts) inside of two forms to create walls.
  See: **Nk'Mip Desert Cultural Centre,** British Columbia - by: HBBH Architects
  http://www.archdaily.com/10629/nkmip-desert-cultural-centre-hbbh-architects

Rammed Earth Works
11 Basalt Road
Napa, CA 94558-3929
Phone: 707.224.2532

http://www.rammedearthworks.com/

*"**Rammed Earth Works** installs rammed earth walls for select clients and leverages a global reputation to promote research, refine technologies, disseminate information and build the credibility that makes rammed earth one of the most environmentally attractive building systems."*

One last ***Rammed Earth*** link . . .
https://sustainablebuildingdesign.wordpress.com/2014/09/02/the-beauty-of-rammed-earth/

In addition to the above, there are a few other ***Alternative Construction*** options, that are not in the above **http://www.greenhomebuilding.com/** website.

- **Pneumatically Impacted Stabilized Earth (PISE)** -
  Pronounced as 'Pee-Zay', is similar to the process of spaying Gunite or Shotcrete used in making in-ground plaster swimming pools. However, PISE sprays a cement and Earth mixture to one side of a form to create the interior and/or exterior walls. **https://www.youtube.com/watch?v=3dwtn3bE3P8** 3:19

- **Rubble Homes** -
  And other types of buildings are made from the rubble left from earthquakes. Haiti is one location where *Rubble Homes* are being built as a result of the 2010 7.0M disastrous earthquake. Many volunteers from churches

have helped to rebuild using local rubble, along with the efforts of Kennesaw State University, Kennesaw, GA! **http://engineering.kennesaw.edu/rubblehouse/**

Rubble House Promo Video -
https://www.youtube.com/watch?v=XceX1qgyRhk 1:31

- ## Shipping Container Buildings -

All of what we have is because of the many trucks that transport the products to us directly or to our stores. According to the **World Shipping Council**, in 1955, Malcom P. McLean, a trucking entrepreneur from North Carolina, revolutionized the shipping industry when he began *"transporting entire truck trailers with their cargo still inside."* Learn more at:

http://www.worldshipping.org/about-the-industry/history-of-containerization

Now, as a result of standardizing their dimensions, these incredibly strong shipping containers are being reused for residential and non-residential uses. See:

- *Shipping Container Family Home: Building Blocks in Redwoods*
  https://www.youtube.com/watch?v=5D4WHT2F0JA 9:32

- *Top 10: Shipping Container Offices*
  http://www.containerhomeplans.org/2015/03/top-10-shipping-container-offices/

- ## Underground & Earth Sheltered -

Is not being discussed much in our current Sustainable movement, but such a concept was very popular as a result of the 1973-1974 Oil Embargo that enabled the cost of oil and gas to skyrocket in price. **https://history.state.gov/milestones/1969-1976/oil-embargo**

An Underground home is truly 'underground' by building into a hill or mountain, but can also include abandoned missile silos.

- **http://www.undergroundhousing.com/**
  (Book & Video 24:15)

- *Weird US Underground Missile Silo Home*
  https://www.youtube.com/watch?v=0_liNYkZn_M 7:17

  Their company (link below) was referenced in above website:
  http://www.missilebases.com/

An Earth Sheltered or Earth-Bermed structure is partially protected or integrated into the Earth.

- *Tour an Earth-Bermed House in Upstate New York*
  https://www.youtube.com/watch?v=ZgWTeU7zDGI 7:26
  Host Bob Vila interviews Architect Alan Shope

And, as you know, Underground, Earth Sheltered, and Earth-Bermed construction is not just limited to residential applications. See this nice list of non-residential buildings, categorized by each State, at:

- **http://www.subsurfacebuildings.com/usaundergroundbuildings.html**

The US Department of Energy has some additional information at:

- **http://energy.gov/energysaver/efficient-earth-sheltered-homes**

- **Tiny Homes** -
  As a result of people wanting to simplify their lives while also freeing themselves from large mortgages, it is hard not to recognize the increasing interest in Tiny Homes on wheels, or permanent foundations. The Tiny Homes on wheels offers a relatively new lifestyle, where Tiny Homes with permanent foundations have been found around the world for many years, but have gained a new interest, with the recent help of Sarah Suzanka, Architect and her *'Not So Big House'* website and book. **http://www.notsobighouse.com/**

With interest in Tiny Homes with wheels, Andrew & Gabriella Morrison share their Tiny House enthusiasm and related information at this website: **http://tinyhousebuild.com/**

Andrew Morrison - "hOMe" Tiny House
https://www.youtube.com/watch?v=RSzgh3D7-Q0 28:02

In a 'rabbit trail', unrelated, yet related comment, I had the pleasure to meet Andrew Morrison a few years ago at a weeklong, hands-on StrawBale Workshop that he taught in Meadeville, PA. His website offers a wealth of information along with future workshops. **http://www.strawbale.com/**

You can see *'yours truly'* at this Strawbale.com group event: **https://strawbale.com/store/graduates/group-photo-from-rob-wozniak/**

Tiny Homes: *Some additional information*

The below represent additional Tiny House examples:

- **The Edges' Sustainable Tiny House**
  https://www.youtube.com/watch?v=TRHcygdYjBw 15:48

- **Living with a Twist:** (Dumpster (Tiny) Home)
  https://www.youtube.com/watch?v=TvSZL4eppTQ 1:35

And, again . . . in as an additional related, yet unrelated comment, the below website provides a list of Home Building Schools in the United States that promote the above.
http://www.dirtcheapbuilder.com/Home_Building_Schools.htm

Ok. Now that you have reviewed the above Alternative Construction / Alternative Materials, and various resources, how will you integrate such into one of your next projects?!

Getting back on track . . . Please note, from this point forward, the following **Division 13 CSI Sections** and related information pulls us back to the CSI official topics.

13 10 10 - Security Protection

In this day and age, security has become an important topic. With the use of building camera's, and satellite imaging, George Orwell's book *1984* - in many ways has become a reality . . . with the good and bad of *'Big Brother'* watching.

For a selection of **13 10 10** related LEED products, see:
http://www.arcat.com/divs/sec/sec131010.shtml

13 11 00 - Swimming Pools

This section addresses various above and below grade swimming pools and related cleaning equipment components. It would be important to review Building Code requirements:

- The International Swimming Pool & Spa Code (ISPSC) . . . yes, there is such a code ☺. Check it out.

- International Residential Code (IRC):

- **Chapter 42** - *Swimming Pools*

- The International Building Code (IBC):

- **Chapter 10** - *Means of Egress*
- **Chapter 11** - *Accessibility*
- **Chapter 29** - *Plumbing Systems*

Use website to access the *I-Codes, PremiumACCESS or the Free PublicACCESS*: **https://codes.iccsafe.org/**

- **National Plasters Council, Inc.**

1000 North. Rand Road, Suite 214
Wauconda, IL 60084
Phone: 847.416.7272
Fax: 847.526.3993

http://www.npconline.org/

"For more than 25 years the NPC has been the utmost authority for in-ground swimming pool and spa cementitious interior pool surface applicators, material manufacturers, suppliers and distributors. Be sure to visit our site to find the latest news, technical updates, and events impacting the pool industry."

- **The Association of Pool & Spa Professionals**

2111 Eisenhower Avenue
Alexandria, VA 22314
Phone: 703.838.0083
Fax: 703.549.0493

http://apsp.org/

*"Welcome to the official website of the **Association of Pool & Spa Professionals,** the world's largest and oldest association representing the swimming pool, spa and hot top industry."*

For a selection of **13 11 00** related LEED products, see:
http://www.arcat.com/divs/sec/sec131100.shtml

- *See Section **13 34 13 - Glazed Structures,** for related Swimming Pool Enclosure information.*

13 12 00 - Fountains

This section addresses exterior fountains and interior waterwalls and waterfalls.

- **Bluworld of Water**

3093 Caruso Court
Orlando, FL 32806

Toll Free: 888.499.5433
Phone: 407.426.7674
Fax: 407.426.7721

http://www.bluworldusa.com/

*"**Bluworld** was created with the goal of being a trusted "single source" that design professionals could rely on to ensure that their water feature projects would be a success. In years past, designing a water feature, particularly an indoor waterfall, into a space was a risky proposition. Without a "water feature specialist" involved, many indoor waterfalls, and other types of water features, either never worked correctly or were plagued with operational issues. There are many nuances in controlling water that can easily be overlooked or unanticipated even by very component design professionals."*

For additional **13 12 00** related LEED products, see:
http://www.arcat.com/divs/sec/sec131200.shtml

13 13 00 - Aquariums

This Section includes the smaller room size aquariums to the large marine park size aquariums and special aquatic architectural applications.

- **Association of Zoos and Aquariums (AZA)**
  8403 Colesville Road, Ste. 710
  Silver Spring, MD 20910-3314
  Phone: 301.562.0777
  Fax:301.562.0888

  https://www.aza.org/

  "AZA is a 501(c)3 non-profit organization dedicated to the advancement of zoos and aquariums in the areas of conservation, education, science, and recreation. AZA represents more than 200 institutions which meet the highest standards in animal care, provide a fun and educational family experience, and dedicate millions of dollars to scientific research, conservation, and education programs."

- **Reynolds Polymer Technology, Inc.**
  607 Holingsworth Street
  Grand Junction, CO 81505
  Phone: 907.241.4700

  http://www.reynoldspolymer.com/

  "Reynolds Polymer Technology (RPT) is the exclusive manufacturer, fabricator, designer, and installer of R-Cast® acrylic and resin-based products. During our 30 years in business, RPT has completed more than 1,900 projects in 57 countries, including the AquaDom in Berlin, Shark Reef in Las Vegas, and many other custom works."

For additional **13 13 00** related LEED products, see:
http://www.arcat.com/divs/sec/sec131300.shtml

13 17 00 - Tubs & Pools

This Section includes **Hot Tubs; Whirlpool Tubs; Therapeutic Tubs** and **'endless' pools.**

This Section is not intended for bathroom tubs, although there are some listed at the below **13 17 00** website link. All bathroom tubs would ultimately be within Divison 22 - *Plumbing*.

For a selection of **13 17 00** related LEED products, see:
http://www.arcat.com/divs/sec/sec131700.shtml

- See CSI Section ***13 11 00 - Swimming Pools,*** *for related products and information, including code references.*

13 18 00 - Ice Rinks

- **Serving The American Rinks (STAR)**

  1775 Bob Johnson Drive
  Colorado Springs, CO 80906
  Phone: 719.538.1149
  Fax: 719.538.1160

  http://www.starrinks.com/

  *"**Serving The American Rinks (STAR)** is a non-profit national membership organization for individuals, facilities and vendors in the ice rink and arena industry. STAR was formed in 2000 through a joint venture between U.S. Figure Skating and USA Hockey with the intent of servicing the needs of ice skating facilities and their employees in the United States."*

- 

  Ice Rink Engineering and Manufacturing, LLC
  A Division of Multiplex Systems, Inc.
  501 Furman Road, Ste. A
  Greenville, SC 29609
  Phone: 864.232.2591

  http://www.ezglide350.com/contact.php

  *"**Ice Rink Engineering and Manufacturing, LLC** is a family owned and operated business. We've been involved in every aspect of the ice skating industry for well over 60 years."*

  EZ Glide 350 (synthetic Ice-skating surface)
  https://www.youtube.com/watch?v=i2silm8LHKg 8:36

"EZ Glide 350® has revolutionized the synthetic ice skating industry, offering exceptional skating performance for figure and hockey skaters alike. It is durable enough to be used for commercial applications such as hockey training facilities, public skating and traveling ice shows, yet affordable enough to justify residential installations allowing at-home training like never before."

For a selection of **13 18 00** related LEED products, see:
http://www.arcat.com/divs/sec/sec131800.shtml

13 19 00 - Kennels & Animal Shelters

- **The American Society for the Prevention of Cruelty to Animals (ASPCA)**
  424 E. 92nd Street
  New York, NY 10128-6804
  Toll Free: 888.666.2279
  Phone: 212.876.7700

  https://www.aspca.org/

  *"**The American Society for the Prevention of Cruelty to Animals®** (ASPCA®) was the first humane society to be established in North America and is, today, one of the largest in the world."*

For a selection of **13 19 00** related LEED products, see:
http://www.arcat.com/divs/sec/sec131900.shtml

13 20 00 - Special Purpose Rooms

This section includes Clean Rooms; Cold Storage Rooms; Constant Temperature Rooms; Wine Cellars; and Sound-Conditioned Rooms

- **American Cleanroom Systems**
  29722 Avenida de las Banderas
  Rancho Santa Margarita, CA 92688
  Phone: 949.589.5656
  Fax: 949.589.6109

  http://www.americancleanrooms.com/

  *"**American Cleanroom Systems** (are) experts in Pharmaceutical, Medical Device & Industrial Cleanrooms"*

- **AES Clean Technology, Inc.**
  422 Stump Road
  Montgomeryville, PA 18936
  Phone: 215.393.6810

  http://www.aesclean.com/

*"**AES** Lean Construction Methodology for design, manufacture, build and commissioning of modular cleanrooms will deliver benefits to your project by controlling cost, eliminating duplicate efforts and accelerating the project schedule."*

- **Terra Universal, Inc.**
  800 S. Raymond Ave.
  Fullerton, CA 92831
  Phone: 714.578.6017

  https://www.terrauniversal.com/

  "Designing building and equipping critical environments since 1976."

For additional **13 20 00** related LEED products, see:
http://www.arcat.com/divs/sec/sec132000.shtml

13 22 00 - Office Shelters & Booths

This section includes the Guard House and Attendant Booths, Picnic Shelters, and Smoking Shelters.

For a selection of **13 22 00** related LEED products, see:
http://www.arcat.com/divs/sec/sec132200.shtml

*- See CSI Section **13 34 23 - Fabricated Structures,** for related products and information.*

13 26 13 - Storm Shelter Rooms

As a result of natural disasters that ravage our nation, it is wise to build the entire structure, such as a home, to be storm proof. After all, this would be the sustainable way, to protect lives and to prevent rebuilding. In other words, to build, or even to purchase an existing building on a certain site, without considering the potential storm damage is not being wise.

This parable comes to mind . . .

> *"Therefore everyone who hears these words of Mine and acts on them, may be compared to a wise man who built his house on the rock. And the rain fell, and the floods came, and the winds blew and slammed*

against that house; and yet it did not fall, for it had been founded on the rock. Everyone who hears these words of Mine and does not act on them, will be like a foolish man who built his house on the sand. The rain fell, and the floods came, and the winds blew and slammed against that house; and it fell—and great was its fall." Matthew 7: 24-27

If not constructed properly and in an unsustainable way, consider the potential loss of life and all the debris that is to be cleaned up and disposed of. This is where the Building Codes come in. They are as a result of learning from past disasters, along with much product and material testing so to prevent such from happening again. The Codes are minimum requirements though, with the intent to protect the **Health**, **Safety** and **Welfare** for the occupants.

But, if it is not practically or financially possible to make the new structure *more* storm proof, (going *beyond* the code), or using alternative materials that are fire, tornado, and/or hurricane proof . . . or an alternative geographical location . . . or, if it is an existing building, . . . then it would be wise to provide a **Storm Shelter / Safe Room** - at the very least.

- **Federal Emergency Management Association (FEMA) - Safe Rooms**
  26 Federal Plaza, #1337
  New York, NY 10278
  Phone: 212.680.3609

  https://www.fema.gov/safe-rooms

  FEMA Disaster Recovery Center
  1904 Surf Avenue
  Brooklyn, NY 11224
  Toll Free: 800.621.3362

  http://www.fema.gov/

- **F5 Storm Shelters & Safe Rooms**
  348 Delmus McMurry Road
  Baskin, LA 71219
  Phone: 318.248.2994

  http://www.f-5stormshelters.com/

  "Our shelters are ideal for churches, fire departments, town halls, schools, etc. We now build large shelters for businesses and communities. Our safe rooms can go inside any home or building that is going to be built, or outside existing homes or businesses."

- **U.S. Storm Shelters**
  Mailing Address:
  P.O. Box 1324
  Decatur, TX 76234

Physical Address:
443 Private Road 4215
Decatur, TX 76234
Toll Free: 800.868.5799 / 800.379.9712
Para Espanol: 940.735.7176

http://www.usstormshelters.com/about.php

*"**U.S. Storm Shelters** has three primary products. These include exterior, in-ground concrete storm shelters, and above-ground interior saferooms, both large and small. All of our products are built strong to shield your family or employees from the dangers of extreme weather and tornadoes. Please view our information for concrete shelters and interior safe rooms, then contact us if you would like to discuss how U.S. Storm Shelters can help keep you safe from the storm."*

13 27 00 - Vaults

Safe Rooms should not be confused with vaults. A Safe Room typically accommodates people. A vault is for possessions.

For a selection of **13 27 00** related LEED products, see the following:
http://www.arcat.com/divs/sec/sec132700.shtml

13 31 00 - Fabric Structures

Fabric Structures offer great and unique opportunities and are categorized under the main CSI 13 30 00 - Special Structures section. They can be broken down into two categories: **Air-Supported** (or Air-Inflated) and **Tensile Structures**. In my mind, they could even be added to the unique or otherwise Alternative Construction list found at the beginning of Division 13. But, none the less, such advancements in **Tensioned Membrane Structures** have allowed for incredible clear-span structures, such as Sports Stadiums, Amphitheatres, Field Houses and airports, (such as Denver International Airport), that offer a soft, artistic aesthetic. Potentially someday, the technology and costs will allow for fabric to be used for residential roofs as well.

On a personal note, being originally from the Buffalo, NY area, I was influenced early on with the presence of BirdAir, Inc.

- **BirdAir, Inc.**
  65 Lawrence Bell Drive, Ste. 100

Amherst, NY 14221
Toll Free: 800.622.2246
Phone: 716.633.9500
Fax: 716.633.9850

http://www.birdair.com/tensile-architecture/history

"Today, sustainable design is an important topic when designing new building construction. Architects and designers are always searching for architectural environmentally friendly products to help create efficient and sustainable building solutions. As a member of the U.S. Green Building Council (USGBC), ***Birdair's*** *tensile membrane structures can help architects/designers achieve and earn Leadership in Energy and Environment Design (LEED™) and Energy Star project certifications and credits for their latest design projects."*

Tensioned Membrane Structures: *Some additional information*

- **Tensotherm™** with LUMIRA® Aerogel
  A partnership between BirdAir, Cabot Corp & Geiger Engineers

  http://www.birdair.com/press/birdair-inc-partners-cabot-corporation-and-geiger-engineers-offer-tensothermtm-breakthrough

- See CSI Section ***07 42 63 - Fabricated Wall Panel Assemblies,*** *for related Aerogel products and information.*

13 32 00 - Space Frames

A **Space Frame** is commonly a light-weight skeletal structure, that is made up of an inverted pyramids connected together by steel struts, to form a large clear-span roof. Like the *CSI Section* ***13 31 00 - Fabric Structures***, Space Frames are not too common, and again, in my mind they could be added to the unique or otherwise Alternative Construction list found at the beginning of Division 13.

- **Delta Structures, Inc.**
  811 Eagle Drive
  Bensenville, IL 60106
  Phone: 630.694.8700
  Fax: 630.694.8787

  http://www.deltastructures.com/

  *"**Delta Structures, Inc.** designs use significantly less material (~60%), are lighter weight and architecturally pleasing, (while creating) a small carbon footprint. (The Space Frame) Structures are made from recycled material and are, themselves, fully recyclable. Source material is abundant and locally available eliminating the need for scarce, high cost material. Significantly less energy consumption is involved in the fabrication of space frame structures. Less energy*

is required to produce the smaller material mass; structures are lighter in weight and are fabricated and delivered in compact unit sizes."

*"In 2000, **Delta Structures, Inc.** acquired the **Unistrut** Space Frames Division of Allied Tube and Conduit Corporation (a TYCO International company) allowing Delta Structures, Inc. to meet market needs in greater measure."*

13 33 00 - Geodesic Structures

In the late 1940's, American Architect, **Buckminister Fuller** is credited in popularizing **Geodesic Domes**. Again, a unique and creative concept that had hoped to improve and provide human shelter.

- **Buckminister Fuller Institute**
  181 N. 11th Street
  Brooklyn, NY 11211
  Phone: 718.290.9280

  https://bfi.org/about-fuller/big-ideas/geodesic-domes

Over the years, others have improved on the Geodesic Dome concept, making them not only efficient to build, but also efficient to operate, as well as storm proof. This appears to be true sustainable structure!

To this day, (on a personal note), I only had one opportunity to be in an existing Geodesic Dome for a client. The challenge is that walls are not vertical and the floor plan is not rectangular. But, the client was happy with the drawings for the proposed alterations, but unfortunately, I do not know if the project was carried out.

- **American Ingenuity (Ai)**
  8777 Holiday Springs Road
  Rocklege, FL 32955-5805
  Phone: 321.639.8777

  http://aidomes.com/

  *"For over 40 years **American Ingenuity (Ai)** has been manufacturing Geodesic Dome Kits with prefab panels for homes or businesses (that can withstand up to) 225 mph Wind & F4 Tornado Warranty. Fire Resistant Concrete Exterior. Save 50%-60% on cooling & heating costs."*

- **Natural Spaces Domes**
  37955 Bridge Road
  North Branch, MN 55056
  Toll Free: 800.733.7107
  Phone: 651.674.4292

  http://www.naturalspacesdomes.com/

  Natural Spaces Domes *has been building "green" geodesic dome homes since we started building domes in the 1970's. Some of* *our domes are already equipped with solar panels, wind generators and are so efficient (with R-values that practically go through the roof) that they produce more energy than they consume. A **Natural Spaces Dome** is uniquely designed for extremely high winds and earthquakes."*

13 34 13 - Glazed Structures

This Section includes **Greenhouses, Solariums and Swimming Pool Enclosures.**

- ### Libart Retractable Roofs & Enclosures

  600 Defiance Avenue
  Hicksville, OH 43526
  Phone: 419.542.0247
  Fax: 419.542.0255

  https://www.libartusa.com/pool-enclosures/

  "The **Libart Retractable Roofs & Enclosures** provides *"increased usability . . . save on maintenance costs . . . (with a) wonderful aesthetic appeal, (for) Restaurants, Hotels and Homes".*

- ### Renaissance Conservatories

  132 Ashmore Drive
  Leola, PA 17540
  Toll Free: 800.882.4657
  Fax: 717.661.7727

  http://www.renaissance-online.com/

  *"**Renaissance Conservatories** is a team of conservatory design professionals dedicated to perfecting the art and architecture of designing, manufacturing, building and servicing exquisite glass conservatories,* (along with) *Garden Windows & Garden Bay Windows; Lanterns & Skylights; Commercial Horticultural Greenhouses, Window & Door Systems; and Pool & Spa Enclosures."*

- ### Solar Innovations, Inc.

  31 Roberts Road
  Pine Grove, PA 17963
  Toll Free: 800.618.0669
  Phone: 570.915.1500
  Fax: 570.915.6083 or 800.618.0743

  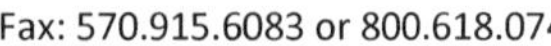

  http://www.solarinnovations.com/folding-glass-walls/

  *"**Solar Innovations, Inc.** is a single source provider of residential and commercial folding, tilting, sliding, and stacking glass doors, walls, windows, and screens; skylights; greenhouses; conservatories; curtain walls; and more. If you don't see the product you're looking for, please contact your Solar Innovations, Inc. sales representative at: 800-915-1500."*

*- See CSI Sections **08 80 00** - **Glazing**, and **13 11 00** - **Swimming Pools**, for related products and information.*

13 34 16 - Grandstands & Bleachers

For a selection of **13 34 16** related LEED products, see:
http://www.arcat.com/divs/sec/sec133416.shtml

13 34 18 - Post Framed Building Systems

Post Framed Buildings, sometimes known as Pole Barns, offer an alternative engineered building type that uses solid sawn posts for its framing, in lieu of other typical materials and systems.

- **National Frame Building Association** (NFBA)
  8735 W. Higgins Road, Ste. 300
  Chicago, IL 60631
  Toll Free: 800.557.6957
  Fax: 847.375.6495

  http://www.nfba.org/About/default/whoweare.html

  *"**NFBA** is the only national trade association that represents post-frame industry professionals. The association exists to support its members and stimulate the growth of the post-frame industry. To do so, the association helps its members by providing industry tools and code resources, education, access to technical and legal experts, builder accreditation programs, newsletters and post-frame market development updates, and networking opportunities."*

- **Morton Buildings, Inc.**
  252 W. Adams Street
  Morton, IL 61550-1804

  Toll Free: 800.447.7436
  Fax: 309.263.6341

  https://mortonbuildings.com/

  Morton Buildings designs and manufactures metal framed and sheathed clear-span "*Residential, Farm, Equestrian, Commercial & Community*" Buildings.

13 34 19 - Metal Building Systems

The main difference between a Post Framed Building with metal sheathing, and a Metal Building with metal sheathing (or other exterior cladding), is its 'bent' steel structural framing. See the below Star Building Systems website *(Metal Building 101)*, to learn more:

- **Star Building Systems**
  8600 South Interstate 35 Service Road
  Oklahoma City, OK 73149
  Toll Free: 800.879.7827

Phone: 405.636.2010
Fax: 405.636.2095

http://www.starbuildings.com/rc_metalbuildings101.html

"Over the last few years and continuing today and into the future, green, along with the terms sustainable development and building green, have become common nomenclature. Star Building Systems has and will continue to provide sustainable products to our customers and will strive to keep you abreast of where influential codes and standards are driving the market."

For additional **13 34 19** related LEED products, see:
http://www.arcat.com/divs/sec/sec133419.shtml

13 34 23 - Fabricated Structures

This section includes those buildings that have been pre-fabricated at a manufacturing facility, in a controlled environment, so to not be affected by weather. Typically, the various panels and systems are designed and perfected in a modular, mass-produced way, then shipped by train, to provide a quality product, while maximizing the materials and minimizing labor, to keep costs low.

In the past there have been various successful examples of modular or pre-fabricated buildings that have influenced the companies of today, including:

- Sears, Roebuck & Co. (1908-1940 Pre-Cut, Catalog Kit Homes)

  http://www.searsarchives.com/homes/byimage.htm

- American System-Built Homes: A 1912-1916 partnership with Arthur L. Richards and F.L. Wright - Milwaukee, WI

  https://www.bluffton.edu/homepages/facstaff/sullivanm/wisconsin/milwaukee/wrightprefab/richards.html

- Usonian Automatic Homes: 1936-1950's by F.L. Wright

  http://www.steinerag.com/flw/Artifact%20Pages/PhRtUsonAuto.htm#Exhibit

  Usonian Automatic Example: 1955 Tracy House, Normandy Park, WA

  http://arcadenw.org/article/making-your-own-house-one-block-at-a-time

- *Glide House* by Michelle Kaufmann:
  http://www.prefabs.com/PrefabHomes/MichelleKaufmannDesigns/Gldiehouse.htm

  Related *Glide House* video:
  https://www.youtube.com/watch?v=Fyk53o_RpgE 4:05

- Modular Homes Today Directory:
  Unfortunately this does not appear to include American Systems-Built, Usonian, or other non-traditional styles from the past.

  http://www.modulartoday.com/modular-directory.html

- **Project Frog** - *(Flexible Response to Ongoing Growth)* creates Integrated buildings with modular, configurable parts that are designed to be expanded upon and adaptive.

  http://projectfrog.com/about/faqs

For additional **13 34 23** related LEED products, see:
http://www.arcat.com/divs/sec/sec133423.shtml

*- See CSI Section **13 22 00 - Office Shelters & Booths,** for related modular building products and information.*

13 35 13 - Rammed Earth Walls

*See **Alternative Construction / Alternative Materials** (at beginning of Division 13), for related products and information.*

13 49 00 - Radon Protection

Radon and other forms of radiation are the silent killers that you cannot see or smell. Per EPA, if the levels of radon are at 4 picocuries per liter (pCi/L) or above, it should be addressed for health reasons.

- **US Environmental Protection Agency (EPA)**
  One Potomac Yard
  2777 S. Crystal Drive
  Arlington, VA 22202-3553
  Phone: 202.272.0167

  https://www.epa.gov/radon

As expected, for Radon Mitigation (or Remediation), the Building Code has some minimum requirements. See the following:

- 2009 International Residential Code (IRC), Appendix F - Radon Control Methods
- 2012 International Residential Code (IRC), Appendix F - Radon Control Methods
- 2015 International Residential Code (IRC), Appendix F - Passive Radon Gas Controls

Use website to access the *I-Codes, PremiumACCESS* or *the Free PublicACCESS*: **https://codes.iccsafe.org/**

- Radon Mitigation in New Construction: **https://www.youtube.com/watch?v=zwz4yPgv3XA** 12:14

Radon: *Some additional information*

- American Lung Association Radon Test Kits **http://www.lung.org/local-content/_content-items/support-and-community/local-support/mpor-radon-test-kits.html**
- EPA Radon Zone Map **https://www.epa.gov/radon/find-information-about-local-radon-zones-and-state-contact-information**
- EPA Citizens Guide to Radon **https://www.epa.gov/sites/production/files/2016-02/documents/2012_a_citizens_guide_to_radon.pdf**
- Pennsylvania Radon Levels and other Information **http://pa-radon.info/**
- See CSI ***Section 03 10 00 - Concrete Forming and Accessories*** - for Form-A-Drain, as one Radon Remediation options.

End of Division 13

Space for notes:

DIVISION 14 - Conveying Equipment

This Construction Specifications Institute (CSI) Division includes **Dumbwaiters, Elevators, Escalators, Moving Walks, Lifts** other means to move people or items easily in the workplace, public place or home.

With the above, the **Americans with Disabilities Act** (ADA) of 1990 comes to mind. And, as mentioned in *CSI **Division 02** - **Existing Conditions,*** this is not code, but law - and applies to all non-residential buildings to address the needs of all employees and the public.

When it comes to the owner-occupied residences, if the home was not built originally to have a zero-step entry and other Universal Designed features, the owners may choose to renovate when finances and the structure and space allows. **Universal Design**, sometimes known as **Barrier-Free Design**, (and other names), to ultimately allow **'Aging-in-Place'** so to benefit all people, in all stages of life. The following features provides accommodations for not only the owners but also the visitors, (allowing *'Visitability')*. Such design offers true ***'SustainAbility'*** for all ages. After all, if you have the *greenest* home, and yet because of Birth complications, an Accident, a crippling Disease, or your Age (young or old) prevents you to enter and move around your home, this home is not able to sustain you! "If you cannot live there, it is not Sustain*Able*!"

Even though the ADA is not code, the Building Code, (and *ICC A117.1*) provide the ADA requirements. The following features encourage Aging-in-Place / Universal Design:

- Exterior

- Accessible Entrances (Zero-Step entry)
- Accessible areas around exterior (handrails & level surfaces)
- Improved exterior lighting (for security and to prevent falls)

- Interior

- Grab bars where necessary (including on exterior)
- Non-slip flooring, low-pile carpeting, non-trip transitions
- Stair lifts or elevators
- Door levers and possible offset door hinge hardware
- 36" Light switch / Thermostat wall heights
- Appropriate Light levels (Ambient/General, Task & Accent)
- 24" Wall receptacle heights
- 48" wide hallways (min.)
- 36" wide doors and doorways (32" 'clear' opening min.)
- First Floor Bedroom (Ranch or Cape Cod style home)

- Adaptable & Useable Kitchens (appropriate heights for counters, sinks & appliances)
- Accessible Bathrooms

Since the above features are not required by Code, they should be 'prerequisites' for the ***LEED for Homes*** rating system, for a true **Sustain*Able*** environment.

- *USGBC LEED for Homes* examples - (but do they demonstrate Universal Design?) **http://www.usgbc.org/projects/homes**

Conveying Equipment: *Some additional information*

See the International Building Code (IBC) website:
- **Chapter 10** - *Egress*
- **Chapter 11** - *Accessibility*
- **Chapter 12** - *Interior Environment*
- **Chapter 30** - *Elevators & Conveying Systems*

See International Residential Code (IRC), **Chapter 3**:
- Section **R310** - *Emergency Escape & Rescue Openings*
- Section **R311** - *Means of Egress*

. . . And potentially others.

Plus: ***ICC A117.1-2009*** *Accessible and Usable Buildings and Facilities* (found at the ICC Standards tab)

Use website to access the *I-Codes, PremiumACCESS or the Free PublicACCESS*: **https://codes.iccsafe.org/**

*See Divison **22** - **Plumbing,** for potential related products and information, and **Section 02 20 00 - Assessment**, for the ADA 20% Disproportionality Rule, as it relates to Existing Buildings.*

- **Making Historic Properties Accessible**
  National Park Service
  US Department of the Interior
  Technical Preservation Services
  Preservation Brief 32

https://www.nps.gov/tps/how-to-preserve/briefs.htm

Universal Design and Aging-In-Place are both promoted by various organizations, including:

- **National Association of Home Builders** (NAHB)
  Certified Aging-In-Place Specialist (CAPS)
  1201 15th Street NW
  Washington, DC 20005
  Phone: 800-368-5242

  http://www.nahb.org/en/learn/designations/certified-aging-in-place-specialist.aspx

14 10 00 - Dumbwaiters

Dumbwaiters are an old amenity that have made a comeback providing the ability to hoist food, groceries, laundry, boxes or other items from one level to another. In many instances the companies that manufacture dumbwaiters also produce elevator and other conveying equipment.

- **Butler Buddy**
  1665 Spring Garden Avenue
  Berwick, PA 18603
  Toll Free: 888.441.9810
  Phone: 570.759.0550
  Fax: 570.759.8015

  http://www.butlersbuddy.com/

  *"**Butlers Buddy** manufactures a complete line of standard and custom residential and industrial dumb waiters for home or industry to suit your needs. Our unique patented electric dumb waiter design reduces approximately 80% of labor normally associated with a product of this type including manual dumb waiters."*

- **Access Elevator & Lifts, Inc. (AEL)**
  Headquarters
  951 S. Saddle Creek Road
  Omaha, NE 68106
  Toll Free: 800.397.4000
  Phone: 402.553.7000

  http://www.accesselevatorinc.com/

  *"Whether you're a school needing to make your gym or stage ADA-compliant, or a homeowner who just wants to see his basement, **AEL** has a stair lift, wheelchair lift, elevator, or portable lift that's right for you. We also offer and install best-in-class lifts from the most respected industry manufacturers."*

For additional **14 10 00** related LEED products, see:
http://www.arcat.com/divs/sec/sec141000.shtml

*- See CSI Section **14 91 00 - Facility Chutes,** for related products and information.*

14 20 00 - Elevators

As you know, since it is less expensive to build upwards, (because of less land, and a smaller building footprint) elevators have enabled buildings to create and modify our various skylines. And, no longer just a convenience, such a conveying system is required by the Americans with Disabilities Act as discussed at the beginning of Division 14.

Over the years, some of the commercial elevator companies have dominated the market, and with it, it is worth touching on the topic of **proprietary** vs. **non-proprietary** elevators systems. Building owners and their building design team should research not only first cost of the elevator installation, but also understand how regular maintenance and emergency repairs will be handled, including where geographically the technician is based out of? The name brand is good, but once a proprietary system is in place, typically only this company can make such repairs, and the building owner is then at the mercy of such a technician and related costs. A non-proprietary, independent elevator company can be based locally and can install and maintain a variety of open-market elevators and their equipment, which can offer flexibility and options through the life of the elevator. As with anything, it would be important to do research for what would be the best choice. See this website, along with other personal/professional research.

http://colleyelevator.blogspot.com/2014/08/proprietary-vs-nonproprietary-elevator.html

Some manufactures are:

- **Otis Elevator Company**
  Headquarters
  10 Farm Springs Road
  Farmington, CT 06032
  Toll Free: 888.441.6847
  Phone: 860.676.6000

http://www.otis.com/site/us/Pages/default.aspx?menuId=1

"Otis Elevator Company is the world's largest manufacturer and maintainer of people-moving products, including elevators, escalators and moving walkways - a constant, reliable name for more than 160 years.

(By mid-2013), Otis' new facility in Florence, South Carolina, will become a manufacturing center of excellence for the United States and Canada."

- ## Port Elevator, Inc.

  Main Office:
  Port Elevator - Lycoming County
  941 Nichols Place
  Williamsport, PA 17701

  Southern PA Office:
  Port Elevator - Dauphin County
  7951 State Route 209
  Williamstown, PA 17098

  Toll Free: 800.326.8422
  Phone: 570.326.4446
  Fax: 570.326.9009

  *"**Port Elevator Inc**. is the largest independently owned elevator company in central Pennsylvania. We install only open source, non-proprietary domestically manufactured elevators."*

- ## Vertical Systems, Inc.

  690 11th Street, NW
  Atlanta, GA 30318
  Phone: 404.581.0094
  Fax: 404.581.0834

  *"**Vertical Systems** ensures that our customers enjoy the very best of modern elevator service and technology including:*
  - *Full Range of Services*
  - *Quick Response Capabilities*
  - *One-to-One Customer Care*
  - *Applied Technical Expertise*
  - *Extensive Resources"*

With the retiring Baby Boomer's (born 1946 - 1964), **Residential Elevators** are becoming more and more popular as a way to enable Aging-in-Place, and potential in-home care. As I witness various loved ones around me, (including myself) age, I am in total favor of such. I personally recall seeing a simple, cage-like, (original equipment) elevator in an early 1900's modest home - that I happen to have a photo of. So, this is not a new concept by any stretch of the imagination! Please see the related discussion at the beginning of Division 14.

Some manufacturers are:

- ## Inclinator Company of America

  601 Gibson Boulevard
  Harrisburg, PA 17104
  Phone: 717.939.8420

  http://www.inclinator.com/

  *"America's favorite and most trusted residential elevator company understands the value of staying in your home - that's why **Inclinator** is the only residential elevator manufacturer that offers fully customizable personal elevator solutions to fit any space. Be in control of your lifestyle and quality of life with Inclinator, and enjoy many more years in your home."*

- **Residential Elevators, Inc.** (REI)
  Corporate Office & Manufacturing Office:
  20 Residential Drive
  Crawfordville, FL 32327

  Mailing Address:
  2910 Kerry Forest Parkway, D4-1
  Tallahassee, FL 32309

  Toll Free: 800.832.2004
  Phone: 850.906.3054
  Fax: 850.926-5319

  https://www.residentialelevators.com/index.htm

  *"(**Residential Elevators, Inc.** offer) Factory Direct installation of Home Elevators throughout the Southeast United States, Eastern Seaboard, Texas and California - (so to) Enjoy commercial elevator features at residential elevator prices."*

*- See CSI Section **14 10 00 - Dumbwaiters,** for related products and information.*

Elevators: *Some additional information*

- **SnapCab** Paneling Systems Simplified
  US Headquarters & Manufacturing
  175 Titus Avenue
  Warrington, PA 18976
  Toll Free: 888.766.7834
  Fax: 877.855.3714

  http://www.snapcab.com/

  *"**SnapCab** patented interlocking panel system is fire rated, elevator code and LEED compliant (providing) custom results - without custom cost or aggravation."*

  "Working with us is easy. From design sessions to our state-of-the-art showroom, sample materials and design boards, to our huge catalog of options and custom-designed solutions, you can specify your project in no time."

14 30 00 - Escalators and Moving Walks

Obviously, a great convenience, both vertically and horizontally. And, totally appreciated at *Phoenix Sky Harbor Terminal 4*, and other large airport complexes, as you strive to make the next flight. It is amazing how just this little extra boost of speed helps. Do you agree?! But, at what expense?!

Now, during the late hours, those Moving Walks and Escalators many times are still moving. So, is there a way to have a motion sensor put the unit into sleep mode when not in use? Well, here is a report, sponsored by the *Federal Aviation Administration*, that aims to do such:
http://onlinepubs.trb.org/onlinepubs/acrp/acrp_rpt_117.pdf

I find it interesting though that Thomas Edison filmed an outdoor ***Moving Sidewalk Paris Expo 1900***! So, as with residential elevators, moving sidewalks are not a new concept by any stretch of the imagination!

https://www.youtube.com/watch?v=BjpCVQgKZsc 1:48

Now fast-forward in time to the *Jetsons* and find all sorts of moving sidewalks, treadmills and conveying systems for use. It appears that this 1962-1963 animated futuristic lifestyle is slowly becoming a reality.

https://www.youtube.com/watch?v=t2Z8kPpLg1g 2:10

For a selection of **14 30 00** related LEED products, see:

http://www.arcat.com/divs/sec/sec143000.shtml

14 40 00 - Lifts

This section addresses not only lifting people (like an elevator, without being in an enclosed elevator cab), but also lifting other things, like vehicles. Take a look!

As with moving sidewalks, car-elevators are not new either. At the website below, you can view *"Footage of the first car-elevator parking garage, downtown Chicago, 1932.":*

http://twentytwowords.com/footage-of-the-first-car-elevator-parking-garage-downtown-chicago-1932/ 1:05

Now, fast-forward to present day, and here is an Automated Parking Garage at *1706 Rittenhouse Residential Tower* project, Philadelphia, PA.

https://www.youtube.com/watch?v=w3vtpGtyw1k 1:38

Since it is less expensive to build up, or down - most typically in an urban environment where land goes for a premium price, this parking concept is a 'no-brainer', while also reducing code related ventilation requirements within.

Vertical Platform Lifts and **Stairlifts**, to accommodate residential and non-residential accessibility for people, should be considered . . .

- **Bruno**

1780 Executive Drive
Oconomowoc, WI 53066
Phone: 262.567.4990

http://www.bruno.com/store/vertical-platform-lift/

Bruno manufactures Stairlifts, Scooter & Powerchair Lifts, Vertical Platform Lifts, and Valet Signature Seating products. Contact **Bruno** for more information, including dealer locations.

- **USA Made stairlift manufacturing companies -** **http://www.stairliftlinks.com/usmanufacturers.htm**

For additional **14 40 00** related LEED products, see:
http://www.arcat.com/divs/sec/sec144000.shtml

14 70 00 - Turntables

For a selection of **14 70 00** related LEED products, see:
http://www.arcat.com/divs/sec/sec147000.shtml

14 80 00 - Scaffolding

This section addresses various scaffolding and platforms used during the construction phase, regular window cleaning of tall buildings, and other uses.

For a selection of **14 80 00** related LEED products, see:
http://www.arcat.com/divs/sec/sec148000.shtml

14 91 00 - Facility Chutes

A chute *(or shaft, per the Building Code)* designed for **laundry, garbage, recycling,** and/or other items can be found in some multi-story single-family homes, but are even more common in various high-rise buildings. These chutes in non-residential buildings, ultimately require fire-rating, and shall be a smooth lined shaft that allows gravity to do its job, while not harboring odors, or the chute can be equipped with an Odor Control Service.

Laundry Chutes are a great benefit for a multi-story, or even a ranch style home with a basement. Beyond using it for laundry, (as a sheet metal lined chase builder amenity, installed between studs and equipped with a small metal door on the bathroom wall) . . . I personally recall growing up, using one as an intercom device! ☺

For non-residential applications, see related International Building Code (IBC), **Chapter 7** - *Fire & Smoke Protection Features,* specifically Sections related to *Shaft* and *Shaft Enclosures.*

Some manufacturers are:

- **Chutes International**
  33 Industrial Park Drive
  Waldorf, MD 20602
  Toll Free: 800.88.CHUTES
  Phone: 240.448.5000
  Fax: 301.753.4108

  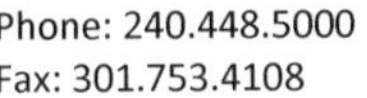

  http://chutes.com/our-products/

  "We Set The Standard. Compare our chutes – Including the Quality, Options, Prices, and Service Record – With ANY in the country! You'll see why CHUTES International is the clear leader in the industry. ***CHUTES*** *can design and build, exclusively furnish, or furnish and install our complete line of internal trash & linen chutes, trash or debris chutes, trash compactors, containers, and external steel and plastic debris chutes to suit your budget."*

- **Midland Chutes**
  9521 W. Ainslie Street
  Schiller Park, IL 60176
  Phone: 847.678.4007
  Fax: 330.677.2252

  http://www.midlandchutes.com/

  "Midland Chutes*, formerly known as Midland Metalcraft Company, was originally incorporated in 1976. Since then, the company has continued to manufacture linen and trash chutes.* ***Midland Chutes*** *is proud of it's 30-year history of meeting the needs of our customers in the Architectural High Rise Industry.* ***Midland Chutes*** *is the largest manufacturer of chutes in the USA.*

For a selection of **14 91 00** related LEED products, see:
http://www.arcat.com/divs/sec/sec149100.shtml

- See CSI Section ***14 10 00 - Dumbwaiters,*** *along with 'doing an internet search for Residential Laundry Chutes', for related products, images, and information.*

End of Division 14

Space for notes:

DIVISION 21 - Fire Suppression

As mentioned previously, the Building Code is a 'design tool', and no differently as it relates to Fire Suppression. Since Fire Suppression, in the form of 'active' Fire Sprinklers, is not required for all buildings, Fire Suppression, in the form of 'passive' building construction is a requirement in today's code. However, with existing structures, this was not the case - and especially true with ***Balloon Frame*** construction. Balloon Framing provided continuous stud cavities from foundation sill to the roof top plate which acts like a fire in a chimney - raging upward to the attic.

The use of **intumescent fire caulk** is used today at even the smallest of penetrations, such as a thermostat wire that penetrates the wall plate, to a small gap around plumbing and other similar applications. The goal is contain a fire as much as possible.

Fire ratings for 'passive' building construction are dependent on topics that include: The **Use &**

Occupancy Classification (also known as Use Groups); the **Building Heights & Areas**; the **Type of Construction**; along with a few other considerations, including **local ordinances**.

So, the Building Codes and related references that come to mind are:

- International Fire Code

- International Building Code (IBC):
  - **Chapter 3** - *Use & Occupancy Classification*
  - **Chapter 5** - *General Building Heights & Areas*
  - **Chapter 6** - *Types of Construction.*
  - **Chapter 7** - *Fire & Smoke Protection Features*
  - **Chapter 8** - *Interior Finishes*
  - **Chapter 9** - *Fire Protection Systems*

- International Residential Code (IRC):
  - **Chapter 6** - *Wall Construction*
  - **Chapter 8** - *Roof-Ceiling Construction*
  - **Chapter 9** - *Roof Assemblies*

Use website to access the *I-Codes, PremiumACCESS or the Free PublicACCESS*:
https://codes.iccsafe.org/

If you were not aware, local ordinances can mandate fire sprinklers including for single-family homes. As noted in *'The Case For Residential Sprinklers'* article below states: *"San Clemente, CA in 1978, was the Nations first jurisdiction to require residential fire sprinklers in all new properties."*
http://www.ips-blazemaster.com/residentialsprinklerscase.html

After this, other cities have followed suit. See the ***Fire Sprinkler Initiative*** website below, for State Fire Sprinkler Requirements:
http://www.firesprinklerinitiative.org/legislation/sprinkler-requirements-by-state.aspx

21 13 80 - Residential Sprinkler Systems

Since Residential Sprinklers are still unfamiliar to most, even now almost 40 years after the 1978 San Clemente ruling, this video shares not only the importance of such, but also shows a step-by-step process of installing a sprinkler system in an existing residence.

Installing a Home Fire Sprinkler System - Ron Hazelton

http://www.ronhazelton.com/projects/installing_a_home_fire_sprinkler_system 8:25

For a selection of **21 13 80** related LEED products, see:
http://www.4specs.com/s/21/21-1380.html

21 20 00 - Fire-Extinguishing Systems

Fire Extinguishers are nice - if you know how to use them. And, if their pressure is not maintained, obviously this is also not good. And, what about if a fully-charged Fire Extinguisher is released onto a kitchen grease fire, this could cause the grease to spread dangerously. So, in this case, a 'passive' method of simply covering the burning pan with a lid or cookie sheet would be the preferred choice. Then, turn off the burner. This video discusses such:

NBC Today Show: Kitchen Fire Safety
https://www.youtube.com/watch?v=6Bvwtr6mdF0 5:38

The next video emphasizes Baking Soda as well as the '*Stove Top Fire Stop Venthood*' product.

Tips Help Prevent, Fight Kitchen Fires
https://www.youtube.com/watch?v=pEVl8R9Q9EM
3:49

- **U.S. Department of Labor**
  Occupational Safety & Health Administration (OSHA)
  200 Constitution Avenue, NW
  Room Number N3626
  Washington, DC. 20210
  Toll Free: 800.321.6742

  https://www.osha.gov/SLTC/etools/evacuation/portable_about.html#class_k

  The above website discusses the Types of Fire Extinguishers, and the basics of Fire Extinguisher Operation.

- **Majestic Fire Protection**
  16134 Valerio Street
  Los Angeles, CA 91406
  Toll Free: 800.918.8978
  Phone: 818.501.4989

  https://majesticfire.com/fire-extinguisher-cabinets.html

  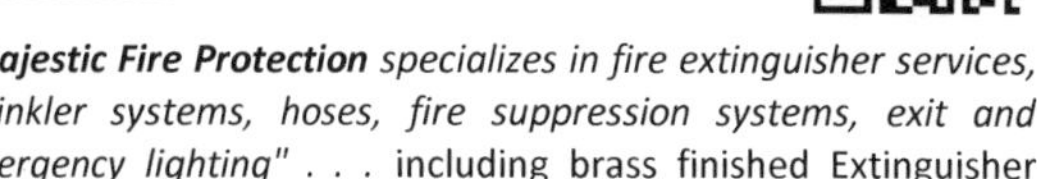

  *"**Majestic Fire Protection** specializes in fire extinguisher services, sprinkler systems, hoses, fire suppression systems, exit and emergency lighting"* . . . including brass finished Extinguisher

cabinets that are more appropriate for the home, especially with a dark bronze acrylic panel.

21 41 00 - Storage Tanks for Fire Suppression Water

If the building is or will be connected to the municipal water supply, this alone is generally reliable. However, to have a main water line break down the street at the same time that a building would catch fire would not be good! This deals with our Nation's unfortunate infrastructure topic. *(See CSI Section **22 13 00** for related information).* Water storage tanks are a wonderful backup plan or alternative for anticipating the need of Fire Suppression, and can be placed under ground. But, if there is a lake, pond, or even a pool on the property would make a perfect emergency water source - as such could be connected with an appropriate pump and hose.

- **Zurn Industries, LLC**
  1801 Pittsburgh Avenue
  Erie, PA 16502
  Toll Free: 855.663.9876
  http://www.zurn.com/

Zurn Industries offers a whole host of products including those for: Pumps, Drainage, Carriers, Interceptors, Hydrants, Commercial Fixtures, Flush Valves, Waterless Urinals, Faucets, Showers, Sinks & Basins, along with piping and repair supplies.

It is understood that **Zurn** is the manufacturer of the PEX fittings used in the ***"All-American Home"*** described at the front of the book, built recently by **Anders Lewendal** and crew - in Bozeman, MT

It is understood that F. L. Wright had trusted this method after the brutal massacre and fire at *Taliesin*, Spring Green, WI. When he designed *Taliesin West,* Scottsdale, AZ. - the pool would serve as a water supply to fight any fires.

*See also CSI **Section 10 44 00 - Fire Protection Specialties,** for related products.*

For a selection of **21 41 00** related LEED products, see:
http://www.arcat.com/divs/sec/sec214100.shtml

End of Division 21

Space for notes:

DIVISION 22 - Plumbing

It is obvious, if we did not have water, we would not have plumbing. In the United States, we are blessed with fresh water, and all that goes with it. Not only as an important life sustaining resource, a universal diluent for beverages & industrial processes, but also very important for health, hygiene, *and* recreation . . . be it at the creek, river, pond, lake, pool or interactive fountain. *(See **Division 02 - Existing Conditions** for important water and water remediation topics.)*

The innovative Romans mastered the ability to build miles and miles of aqueducts so to guide fresh water for much of the same reasons found in the US today.

As emphasized in previous CSI Divisions and related Sections, the Building Code is an important reference when it comes to the minimum requirements for Plumbing. The code becomes a **design tool** for this topic, just as much as it does for others. For instance, in a non-residential building, how many lavatory sinks, toilets, urinals, Drinking Fountains, and Service Sinks should someone be planning, and how much space (square footage) of the building should they consume?

Rather than guessing, the code has the answer. Yes, a designer could add additional fixtures, based on the clients requests, but if the client does not request such, why would he/she add more? More fixtures, which in turn means additional square footage requirements, just means more 'overhead' cost to the project. As we know, the designer should always have the best interest of the client in mind, so I hope that you keep reading.

To use the Building Code as a design tool for Residential, *One & Two Family Dwellings*. See . . .

- Use this website for *PremiumACCESS or* free *PublicACCESS* viewing of ***I-Codes*** *and* ***ICC Standards***: **https://codes.iccsafe.org/**

  . . . then select, International Residential Code (IRC): (See ***I-Codes*** **tab** for the following references)

  - **Chapter 3, Section R304 - *Minimum Room Areas***
  - **Chapter 3, Section R306 - *Sanitation***
  - **Chapter 3, Section R307 - *Toilet, Bath, and Shower spaces.***
  - **Chapter 29 - *Water Supply & Distribution***

To use the Building Code as a design tool for Non-Residential Buildings, see . . .

- Use the same *I-Codes website,* . . . then select, International Building Code (IBC): (See ***I-Codes*** **tab** for the following references)

  - ***Chapter 29 - Plumbing Systems*** and specifically ***Section 2902 - Minimum Plumbing Facilities,*** for the number of Plumbing Fixtures.
  - *Section* ***2903*** (IBC 2009) - ***Toilet Room Req'ts***
  - ***Chapter 11 - Accessibility***
  - ***Chapter 12 - Interior Environment***

- ***ICC G3 - 2011*** *(First Printing)* ***Global Guideline for Practical Public Toilet Design*** (Use the ***ICC Standards*** **tab** for ICC G3 - 2011)

And, for *both* Residential & Non-Residential applications, see . . .

- International Plumbing Code (IPC):
  - *Section* ***604 - Design of Building Water Distribution System*** (Use the ***I-Codes*** **tab** for the IPC.)

- International Private Sewage Disposal Code (IPSDC) (See ***ICC Standards* tab** for the IPSDC)

- ***ICC/ANSI A117.1 - 2009***
  - **Chapter 6** - *Plumbing Elements and Facilities*
  (See ***I-Codes* tab** for the ICC/ANSI A117.1 - 2009)

And potentially others.

- Use this website for *PremiumACCESS or* free *PublicACCESS* viewing of ***I-Codes*** *and* ***ICC Standards***: **https://codes.iccsafe.org/**

As mentioned previously in *CSI* ***Division 10 - Specialties,*** the below ***Planning Guide*** is a valuable tool that shares common and efficient restroom configurations.

- ***Bobrick Planning Guide for Accessible Restrooms***: **http://www.bobrick.com/documents/planningguide.pdf**

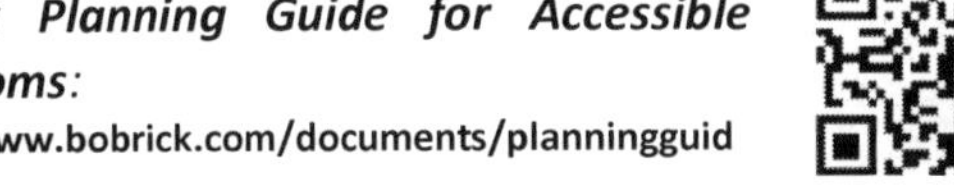

*See **Universal Design** discussion (at beginning of **Division 14**), as well as Section **10 21 13 - Toilet Compartments**, for related products and information.*

*See **Divison 02 - Existing Conditions**, for the ADA 20% Disproportionality Rule, as it relates to Existing Buildings.*

Plumbing: Some additional information
*See heading: **Water That Sustains Us,** found within the Additional Resources section at the end of the book.*

22 10 00 - Plumbing Piping

- **Plumbing Manufacturers International** (PMI)
  1921 Rohlwing Road, Unit G
  Rolling Meadows, IL 60008
  Phone: 847.481.5500
  Fax: 847.481.5501

https://www.safeplumbing.org/health-safety/lead-in-plumbing

PMI offers a wealth of trade related plumbing information, including information about lead in plumbing.

- **What Type of Plumbing Pipe is Best?**
  http://www.benjaminfranklinplumbingiowa.com/what-type-of-plumbing-pipe-is-best/

  The above website communicates the 'pros' and 'cons' of: Copper, Galvanized Steel, PVC, CPVC, PEX, Grey Plastic Polybutylene (PB), and High-Density Polybutylene (HDPE) pipes.

- **Purdue University News Article:**
  Drinking Water Odors, Chemicals above health Standards Caused by 'Green Building' Plumbing
  https://www.purdue.edu/newsroom/releases/2014/Q4/drinking-water-odors,-chemicals-above-health-standards-caused-by-green-building-plumbing.html

- **How to Recognize Different Kinds of Pipes**
  http://www.dummies.com/how-to/content/how-to-recognize-different-types-of-pipes.html

The above *For Dummies* website shares illustrations and descriptions of the:
- **Drain, Waste, Vent (DWV)** common options, including: Cast Iron, ABS, and PVC
- **Potable Water Supply** options: CPVC, PVC, PEX, Steel, and Copper

As mentioned in **Division 02** - *Existing Conditions . . .*

- **Lead from Plumbing Fixtures**
  See the ***Reduction of Lead in Drinking Water Act***. Plumbing fixtures manufactured prior to January 4, 2014 have higher concentrations of lead.
  https://www.iccsafe.org/cs/PMG/Documents/10-31-2013_LeadInPlumbingProducts.pdf

  It is recommended by the EPA to *"Flush water outlets used for drinking or food preparation".*

  http://www2.epa.gov/lead/learn-about-lead#exposed

22 13 00 - Facility Sanitary Sewerage

This Section, includes the **Drain, Waste, & Vent** (DWV) system of a building. **Cast Iron** is one of the options listed above, in *CSI Section **22 10 00 - Plumbing Piping.*** Typically, it was used historically to make up the DWV systems. And, it is slowly making a come-back, as it is recognized for its durability. As noted on the Charlotte Pipe website, their Cast Iron pipe is "*ICC-ES certified as 96% post-consumer recycled content and are 100% recyclable*". And, along with this, it is very quiet. Personally, I recall one of our homes had the DWV pipe in the Dining Room wall was not Cast Iron. So, as a result, every time the upper toilet was flushed during dinner . . . it was an unfortunate sound. ☹

I encourage you to not make this mistake. Although, pipe insulation could be a solution to plastic (PVC) piping, it would require a larger wall cavity.

*(See CSI Section **06 64 00** - regarding the **PVC debate**.)*

- **Charlotte Pipe and Foundry**
  153 Industrial Parkway North
  Muncy, PA 17756
  Phone: 570.546.7666

  And . . .
  2109 Randolph Road
  Charlotte, NC 28207

  http://www.charlottepipe.com/green_building.aspx

"***Charlotte Pipe** has practiced environmental-friendly principles for years, both in our manufacturing processes and with our products. Our cast iron products are ICC-ES certified as 96% post-consumer recycled content and are 100% recyclable. We are also the first manufacturer to offer PVC DWV pipe with recycled content and ReUze, purple CPVC pipe used for non-potable water applications inside of a building.*"

The above website communicates about all the piping Charlotte makes, including: Cast Iron, RePVC® and ReUze®: CPVC.

With the **Sanitary Waste** topic, also brings to mind the condition of the US infrastructure. If you recall, this was one of Mr. Obama's top agenda items in 2008, as he ran for President. One of his plans for *Change*, was to put people back to work, and it was communicated to be similar to *FDR's New Deal,* yet with repairing or rebuilding the infrastructure. Since many of our sewers, waterlines, roads, and bridges are still failing, it appears that this great idea never made much progress. But, thanks to companies like ***Insituform***, such repair costs to water and wastewater systems can be minimized. See below:

- **Insituform Technologies, LLC.**

Global Headquarters
17988 Edison Avenue
St. Louis, MO 63005
Toll Free: 800.234.2992
Fax: 636.519.8010

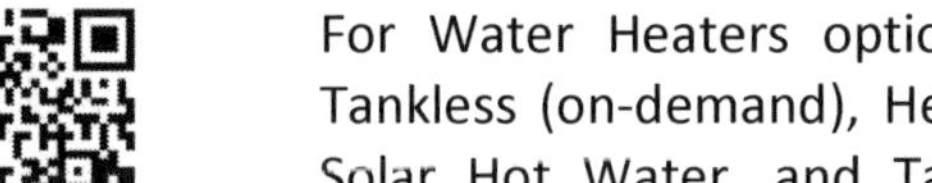

http://www.insituform.com/AboutUs

*"**Insituform is** a leading worldwide provider of cured-in place pipe (CIPP) and other technologies and services for the rehabilitation of water and wastewater pipeline systems.*

*The Company provides cost-effective solutions to remediate operational, health, regulatory and environmental problems resulting from aging and defective pipelines. Insituform's technologies allow its customers to avoid the extraordinary expense and extreme disruption that can result from traditional "dig and replace" methods. Since its formation in 1971, **Insituform** has rehabilitated enough pipe to circle the globe."*

*- See CSI Section **22 10 00 - Plumbing Piping,** for related products and information.*

22 30 00 - Plumbing Equipment

This Section includes - Domestic Water Softeners; Domestic Water Filtration Equipment; Domestic Water Heaters; and Domestic Water Heat Exchangers.

For Water Heaters options, including Conventional, Tankless (on-demand), Heat Pump based, Solar Hot Water, and Tankless Coil and Indirect water heaters, see the **US Department of Energy,** *Energy Saver* website:
http://www.energy.gov/energysaver/water-heating

For a selection of **22 30 00** related LEED products, see:
http://www.arcat.com/divs/sec/sec223000.shtml

22 40 00 - Plumbing Fixtures

To see if restroom (bathroom) fixtures for residential or non-residential buildings meet the **WaterSense®** program requirements, see:
http://www.icc-es-pmg.org/Watersense/

If the fixture is not **WaterSense®** certified**,** the fixtures manufactured today are **'Low Flow'**, to meet the minimum requirements per Code. Some products however exceed the code requirement, but the question is . . . with less and less water, will there be enough water to keep the waste moving in pipes and sanitary sewers?

Use this website for *PremiumACCESS or* free *PublicACCESS* viewing of ***I-Codes** and **ICC Standards***: **https://codes.iccsafe.org/**

Then, access the International Residential Code (IRC):
- **Table P2903.1** - *Required Capacities at Point of Outlet Discharge*

- **Table P2903.2** - *Maximum Flow Rates and Consumption for Plumbing Fixtures & Fixture Fittings* (Compare to International Plumbing Code (IPC), **Table 604.4** - with same title).

And, along with the above, you might consider another way to be water wise. One of my instructors years ago - said it this way:

"If its yellow, let it mellow. If it's brown, flush it down".

Some manufactures are:

- **American Shower & Bath (ASB)**
  A Masco Corporation
  540 Glen Avenue
  Moorestown, NJ 08057
  Phone: 856.235.7700

  http://asbbathingsystems.com/

  ASB manufactures Shower & Bath Ensembles, Bathtub and Shower Walls, Corner Shower Bases, and Utility Sinks

  It is understood that with the creation of Masco Bath, ASB has since closed and/or renamed as Masco. Bath, per Google Maps. See:
  http://www.topix.com/forum/city/savannah-tn/TC08ONUFDBM59NEFG

- **Aqua Glass Corporation**
  A Masco Corporation
  320 Industrial Road
  Adamsville, TN 38310
  Toll Free: 800.632.0911
  Phone: 731.632.0911
  Fax: 800.822.9011

  http://www.aquaglass.com/products/index.html

  Note: It is understood that **Aqua Glass Corp.** is the manufacturer of the shower and tubs used in the ***"All-American Home"***

described at the front of the book, built recently by **Anders Lewendal** and crew - in Bozeman, MT. It is not know if any are Hands-Free or not.

It is understood that with the creation of Masco Bath, this ***Aqua Glass*** *location has closed. See:*
http://www.topix.com/forum/city/savannah-tn/TC08ONUFDBM59NEFG

- ## BioLet Toilet Systems, Inc.

830 West State Street
Newcomerstown, OH 43832
Toll Free: 800.524.6538

http://www.ez-loo.com/

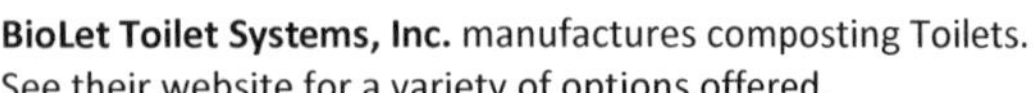

BioLet Toilet Systems, Inc. manufactures composting Toilets. See their website for a variety of options offered.

- ## Delta Faucet Company

A Masco Corporation
Main Office
55 E. 111th Street
Indianapolis, IN 46208
Phone: 317.848.1812

https://www.deltafaucet.com/

Delta Faucet Company manufactures a variety of Kitchen and Bathroom faucets. See their website for a variety of options offered.

- ## Kohler Co.

Headquarters
444 Highland Drive
Kohler, WI 53044
Toll Free: 800.456.4537
Phone: 920.457.4441

http://www.us.kohler.com/us/

The **Kohler Co.** manufacturers a variety of Bathroom, Kitchen fixtures and faucets, along with Bar Sinks and Utility Sinks, and waterless urinals. See their website for a variety of options offered.

Note: It is understood that **Kohler** is the manufacturer of the toilet seats used in the "***All-American Home***" described at the front of the book, built recently by **Anders Lewendal** and crew - in Bozeman, MT.

- ## Mansfield Plumbing Products, LLC

150 East First Street
Perrysville, OH, 44864
Phone: 877-850-3060
Fax: 800-984-7802

http://www.mansfieldplumbing.com/

Mansfield Plumbing manufactures residential and non-residential Sinks, Showers, Shower bases, Tubs, Urinals, Bidets and Toilets, along with their unique *Cascade Rimless Toilet* and *Restore Walk-In Tub*

Note: It is understood that **Mansfield** is the manufacturer of some of the fixtures used in the "***All-American Home***" described at the front of the book, built recently by **Anders Lewendal** and crew - in Bozeman, MT.

- ## Moen, Inc.

Headquarters
25300 Al Moen Driver
North Olmsted, OH 44070
Phone: 440.962.2000

http://www.moen.com/whats-new/innovation/motionsense

Moen, Inc. manufactures MotionSense Hands-Free faucets and other American Made plumbing fixtures, including waterless Urinals.

Note: It is understood that **Moen** is the manufacturer of the faucets used in the "***All-American Home***" described at the front of the book, built recently by **Anders Lewendal** and crew - in Bozeman, MT. It is not know if any faucets installed are *Hands-Free* or not.

- ## Nature's Head Composting Toilet

P.O. Box 250
Van Buren, OH 45889
Phone: 251.295.3043

http://natureshead.net/

*The **Nature's Head** Composting toilet is the most innovative design, offers the best value and the most reliable solution for personal sanitation requirements of any product in the field.*

- ## Toto USA, Inc.

1155 Southern Road
Morrow, GA 30260
Toll Free: 888.295.8134

http://www.totousa.com/

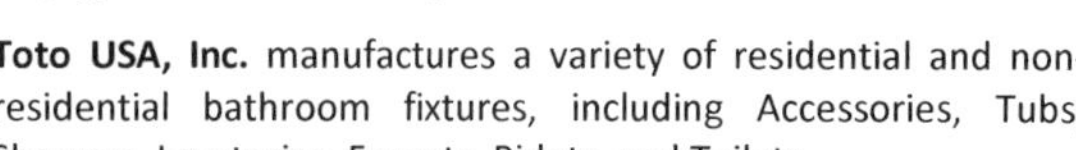

Toto USA, Inc. manufactures a variety of residential and non-residential bathroom fixtures, including Accessories, Tubs, Showers, Lavatories, Faucets, Bidets, and Toilets.

- ## Waterless Co., Inc.

1050 Joshua Way
Vista, CA 92081
Toll Free: 800.244.6364
Phone: 760.727.8823

http://www.waterless.com/

*"For almost a quarter century, **Waterless Co. Inc.** has been the worldwide, industry standard for no-flush/non-water urinals. The simplicity and cost effectiveness of our technology has been appreciated in large, small, busy and outlying facilities."*

- ## Zurn Industries, LLC

1801 Pittsburgh Avenue
Erie, PA 16502
Toll Free: 855.663.9876

http://www.zurn.com/

Zurn Industries offers a whole host of products including those for: Drainage, Carriers, Interceptors, Hydrants, Commercial Fixtures, Flush Valves, Waterless Urinals, Faucets, Showers, Sinks & Basins, along with piping and repair supplies.

It is understood that **Zurn** is the manufacturer of the PEX fittings used in the "***All-American Home***" described at the front of the book, built recently by **Anders Lewendal** and crew - in Bozeman, MT

Plumbing Fixtures: *Some additional information*

*- See CSI Section **11 70 00 - Health Equipment,** for related information on Hands-Free faucets bacteria concerns.*

*See other Division 22 - Plumbing Sections for related products and information, including the topic on **Lead** in **Division 02**, and CSI Section **22 10 00 - Plumbing Piping.***

22 50 00 - Pool and Fountain Plumbing Systems

For a selection of **22 50 00** related LEED products, see:
http://www.arcat.com/divs/sec/sec225000.shtml

*- See CSI Sections **13 11 00 - Swimming Pools,** and **13 12 00 - Fountains,** for related products and information.*

22 60 00 - Gas and Vacuum Systems for Laboratory and Healthcare Facilities

For a selection of **22 60 00** related LEED products, see:
http://www.arcat.com/divs/sec/sec226000.shtml

End of Division 22

Space for notes:

Additional Resources & Thoughts from the Author:

~~~~~~

Employment:

Keep in mind that any of the companies or organizations found within could mean a wonderful summer employment or internship experience for students, and/or long term career opportunity. So, as you are learning about their products & materials, and potentially specifying and/or installing them, you could also be their next valuable team member / employee!

~~~~~~

Made In America -

https://www.youtube.com/watch?v=7AlexLrZ14A 13:30

This episode continues the ***Made in America*** theme that was communicated earlier in the book. See above link for the entire story.

*"At the request of "**World News with Diane Sawyer**," one Dallas family is going to give it a try and share their experiences with the country in a special series, "Made in America," focusing on U.S. manufacturing, jobs and what it all means for the nation's economy. In their single-story home on Snow White Drive, the Usrys lead full and active lives."*

List of materials / products listed in the "*All-American Home*", built by Anders Lewendal, in Bozeman, MT.:
http://abcnews.go.com/images/Business/Made%20in%20America%20List.pdf

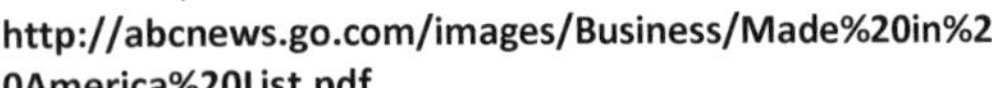

"Thank you for including our efforts in your book. I think we have the same intentions in promoting American made manufacturing.

I applaud your efforts to consider good building science and sustainability when designing and building homes and other structures. Studying the wisdom and word of God ranks up there too.

What I have learned about sustainability and good building science often comes back to diminishing returns. If we, as professionals, seek to build for a higher level of energy efficiency, for example, and fail to understand where those diminishing returns are, we may unintentionally miss an opportunity cost that would provide for better sustainability."

Anders Lewendal

The Greenest Building is the One That is Already Built
Our existing buildings that are able to be maintained for their original purpose, or if because of neglect, renovations may be needed. Either way, our existing buildings have character, scale, textures, and rhythm that sets them apart, creating the community that they are a part of. To remove them would be ignoring such, along with all the embodied energy used to create it in the first place. Do the right thing, Reuse it! Don't dispose of it!
http://www.preservationnation.org/information-center/sustainable-communities/green-lab/lca/The-Greenest-Building_One_Pager.pdf

---

A few additional Resources:

- 4 Specs - An Architectural Library of products & materials
  http://www.4specs.com/
- ARCAT - AEC related resources (CAD, BIM, Specs, etc.)
  http://www.arcat.com
- AUGI - AutoDesk User Group Int'l (Free membership)
  https://www.augi.com/
- Declare - Provides Product Database and 'Nutrition Labels'
  http://declareproducts.com/
- Durability + Design - various AEC related content
  http://www.durabilityanddesign.com/
- **http://www.durabilityanddesign.com/**

  Green Building Supply - 'Your Eco-Friendly Home Center'
- **http://www.greenbuildingsupply.com/**

Some thoughts from the author . . .

As I get older, I tend to marvel at our world, and how it sustains human, animal and plant life throughout each season and through our own 'seasons of life', as expressed in the quote below.

There is a time for everything,
and a season for every activity under the heavens:

a time to be born and a time to die,
a time to plant and a time to uproot,
a time to kill and a time to heal,
a time to tear down and a time to build,
a time to weep and a time to laugh,
a time to mourn and a time to dance,
a time to scatter stones and a time to gather them,
a time to embrace and a time to refrain from embracing,
a time to search and a time to give up,
a time to keep and a time to throw away,
a time to tear and a time to mend,
a time to be silent and a time to speak,
a time to love and a time to hate,
a time for war and a time for peace.

Ecclesiastes, 3:1-8

The above words were put to music by the *Byrds*, and became quite popular many years ago.
See below:

https://www.youtube.com/watch?v=W4ga_M5Zdn4

With this, you might consider . . .

The Earth that sustains us:

Our Earth. However you look at how the Earth came to be, by Creation or Evolution – you need to admit that all the resources that we have been entrusted to manage, are pretty amazing! Is it all by coincidence? Was it by a huge cosmic bang that put things together in such perfect order or is it by intelligent design? By Divine intervention? This is for each of us to answer.

Our buildings that sustain us:

Edward Allen and Joseph Iano, the authors of the book *Fundamentals of Building Construction* state it well when they say: *". . . We build because little that we do can take place outdoors".* (See the beginning of the book for the entire quote.)

Then, not only do we as humans build our various shelters, but animals and insects do too. To co-exist, we are each given various talents and abilities. As we pursue our lives, how and what will we contribute to our World, including the buildings that we are to be good stewards of - during our lifetime?

Food that sustains us:

As the summer months end and the fall comes upon us, I am reminded of the predicable harvest 'bounty', that includes sustainable food for nutrition and health.

> *". . . and the fruit thereof shall be for food, and the leaf thereof for healing."* Ezekiel 47:12

Is it by coincidence that much of our fruit and vegetables are typically not any larger than an adult human hand(s) can hold? These natural foods just so happen to provide the necessary nutrition, while warding off illness! And, is it by coincidence that there are natural cures from the plants around us that scientists and medical doctors keep discovering? Many of these plants were used historically by previous generations, and we are just now either discovering or re-discovering them. And, what about the natural decomposing that takes place, and their various seeds that allow reproduction? Just one of the many examples of this is with apples. The old adage, *"an apple a day keeps the doctor away"* is recognized when one researches apples. Not only are apples good for lung health, but when cut in half, what is revealed almost looks like a pair of lungs! Then, it is noted that apples are not only ready for harvest in the fall prior to the common cold season, but they survive well without decay during these winter months. And, what about the distance between the elbow and mouth to enjoy the apple or any food? Is this all by coincidence?! Or, is it planned sustenance?

Water that sustains us:

As you know, water is the sustainer of all life, for people, animals and plants. Was this also by chance or design? Yet - water can also destroy life - as done by the horrendous damage of a flood or how it can slowly cause damage by the slightest, yet continuous drip.

As we know, water is everywhere. It is in the Earths' subterranean aquifers, the streams, rivers, lakes oceans and clouds. It has been studied through time as **ice, water** and **vapor,** yet it is still H_2O - a marvelous mix of Hydrogen and Oxygen. Some claim glacial melt will raise ocean levels, or that fresh water is disappearing by evaporation or consumption, yet there are those that remind us of the **Law of Conservation of Matter**, the classical physics principle where matter cannot be destroyed – it simply changes states and goes through its cycle - in this case, ice, to water and then to vapor - and back again.

"He covers the sky with clouds;
he supplies the Earth with rain
and makes grass grow on the hills."
Psalm 147:8

So, is there a God that has the days numbered and has established a time for everything - including the water cycle? A planned replenishing?

It is understood that quality food, water and shelter are paramount to sustaining all life, but what about love?

Love that sustains us:

It is very obvious that we as humans do not have any issue following the below Commandment:

> *"Be **fruitful** and **multiply**, and fill the earth, and **subdue it**; and rule over the fish of the sea and over the birds of the sky and over every living thing that moves on the earth."* Genesis 1:28

You might take a moment to see the ***World Population Clock*** tick :
http://www.worldometers.info/world-population/

As the clock ticks, we are called to be good stewards of all that has been entrusted to us individually, as well as with all the resources the Earth provides. Yet, ultimately, if there is a God who claims to be the *Author* and the *Sustainer of Life*, then will He not continue to provide? And, when it is time, will He not follow through with His promise of a New Heaven and a New Earth? Book of Revelation, Chapter 21.

This all ties together when we as humans not only consider important materials and products, but also consider where do the raw materials come from . . . and how selecting such materials and products is a choice. This choice applies to not only those sustainable items used to construct our buildings, but also those for personal use and consumption. As mentioned in Division 02, this *Kia Soul* commercial pretty well sums this up. That is, the importance of making a decision on selecting . . . *'This or That'*.
https://www.youtube.com/watch?v=7geyX1Er1zs 1:00

About the author . . .

The author is originally from the Buffalo, NY area. He currently resides in the Williamsport, PA area and hopes his students, past and present at *Penn College, Vincennes University, Mesa Community College* and *Phoenix Institute of Technology,* along with all the various colleagues he has worked with and met through the years within the AEC profession, along with those that he has not had the pleasure of meeting yet, hopes that they may benefit from this effort that attempts to fill a much needed niche.

He loves people, and is mindful of the fact that we all should strive to be lifetime learners of their Faith and relationships, but also - the environment, design, and construction . . . while being mindful of the consequences and the positive or negative rippling effect of our actions *(or lack of actions).*

Rob is married to the bride of his youth, and has been blessed with three *Hoosier-born* sons. He dedicates this book to God the Father, his parents, his dear wife, family, friends, colleagues, students, and the future of this *great* Country . . . as we are always reminded to learn from the past.

"Blessed is the nation whose God is the Lord". Psm 33:12

"Tis the kindly influence of Christianity we owe that degree of civil freedom, and political and social happiness, which mankind now enjoys, . . . Whenever the pillars of Christianity shall be overthrown, our present republican forums of government - and all blessings which flow from them - must fall with them."

Rev. Jedidiah Morse - Father of American Geography

"Let every student be plainly instructed and earnestly pressed to consider well the main end of his life and studies is to know God and Jesus Christ which is eternal life (John 17:3) and therefore to lay Christ in the bottom as the only foundation of all sound knowledge and learning. And seeing the Lord only giveth wisdom, let every one seriously set himself by prayer in secret to seek it of Him (Proverbs 2, 3). Every one shall so exercise himself in reading the Scriptures twice a day that he shall be ready to give such an account of his proficiency therein."

Harvard 1636 Student Guidelines

"All the scholars are required to live a religious and blameless life according to the rules of God's Word, diligently reading the Holy Scriptures, that fountain of Divine light and truth, and constantly attending all the duties of religion." Yale 1787 Student Guidelines

End of Additional Resources

Index

D

E

F

T

Made in the USA
Middletown, DE
17 December 2018